SOLAR ENERGY

FOR HEATING AND COOLING OF BUILDINGS

Solar Energy
for Heating and Cooling
of Buildings

Arthur R. Patton

NOYES DATA CORPORATION

Park Ridge, New Jersey   **London, England**

1975

Published in the United States of America by
Noyes Data Corporation
Noyes Building, Park Ridge, New Jersey 07656

Foreword

This is Volume No. 7 in our Energy Technology Review Series. Most of the data and other information presented in this book are based on international studies conducted by industrial and engineering firms or university research teams under the auspices of various governments and governmental agencies.

Here are condensed vital data that are scattered and difficult to pull together. Experimental equipment and structures are reviewed and detailed by actual case histories. A short description of commercially available hardware is also included. This condensed information should provide a sound background for action towards combating the energy shortage through the use of solar energy.

Advanced composition and production methods developed by Noyes Data are employed to bring our new durably bound books to the reader in a minimum of time. Special techniques are used to close the gap between "manuscript" and "completed book." Industrial technology is progressing so rapidly that time-honored, conventional typesetting, binding and shipping methods are no longer suitable. Delays in the conventional book publishing cycle have been bypassed to provide the user with an effective and convenient means of reviewing up-to-date information in depth.

The Table of Contents is organized in such a way as to serve as a subject index and provides easy access to the information contained in this book. A valuable list of bibliographic references is also given, followed by an up-to-date collection of names and addresses where one may write for further technical information and specifications of available equipment.

Contents and Subject Index

Introduction

The developing worldwide shortage of petroleum emphasizes the need for alternate energy sources which are both inexpensive and clean. This need is frequently referred to as "the energy crisis." Among possible alternate energy sources the most pollution free, limitless source is solar energy. As with other energy sources, the source exists; the problem is how to tap it. Ultimately, the sun will be harnessed to produce electricity, synthetic fuels and high temperature thermal energy for industrial use; however, the first large scale application of solar energy will likely be to heat and cool buildings since this will require the least number of technological advances from the state-of-the-art.

More than 25% of the total energy used in this country is consumed for the heating and cooling of buildings and the provision of hot water. Therefore, the diversion of this particular energy demand to an alternate source would result in a substantial reduction in the nation's dependence on fossil fuels. The annual incidence of solar energy on buildings in the United States is several times the amount required to heat these buildings; approximately 10^{15} kilowatt hours of solar energy are received on earth annually. It has been projected that by the year 2020 from 25 to 50% of the thermal energy for buildings could be provided from the sun.

Solar radiation had early practical use as a drying agent for foods, in the dehydration of sea water to obtain salts and in the distillation of sea water. It has also been used for some time in Australia, Israel and Japan to heat water. About two to three dozen experimental solar residences have been built in this country and a few experimental units are in operation around the world but no large scale work has been completed and/or documented. The first real support for basic research in the United States for the utilization of solar energy for heating and cooling of buildings came from the Cabot Fund given to MIT and Harvard in the late 1930s. This fund financed pioneering work on home heating,

1

flat plate collectors and photochemical processes. Subsequently, little attention was given by the government or by universities to further this research until the late 1950s. It is reported that government spending from 1950 to 1970 averaged only $100,000 per year for total solar energy research. Since that time, however, interest has increased noticeably as shown by the number of conferences, seminars and symposia held in recent years.

With the onset of the energy crisis a mushrooming interest has been developing in solar and other energy forms and one is exposed almost daily in the news media to the latest research advances. All of this plus greater concern, in the form of funding, by the Federal government, should serve in the near future to draw an increasing number of people, from variously related fields, to the task of improving the technology of solar heating and cooling processes.

In 1974 $1.7 million was allocated for research on solar energy for buildings and facilities and the Energy Research and Development Administration (ERDA) estimates a budget of $4.1 million for 1975 and $21.6 million for 1976. The National Science Foundation (NSF) and the National Aeronautics and Space Administration (NASA) also have funds budgeted for such studies. If these projected amounts materialize, "solar energy" could become a household term, resulting in a "new" industry whose potential as yet can only be imagined.

Perhaps the major problem to be surmounted will be selling solar energy systems to the public. While large first costs for a system will probably be more than compensated for over the life span of a system using relatively "free" energy, there will have to be both an education program and possibly some form of incentive system to induce homeowners and landlords into either new or retrofit installations. Both Indiana and Arizona already offer long-term tax reduction benefits for solar system installations and several other states have legislation pending. Some experts believe that, once the consumer is educated, commercial development could be so rapid that in 20 years the U.S. could be saving 2 million barrels of oil a day.

Solar energy arrives on earth in the form of radiation which is used in basic conversion processes classified into two general groups, those utilizing either solar heat or light. The thermal processes may be classified according to the temperature obtained. Low temperatures are the easiest to obtain, and this is done by means of simple flat plate collectors coated with a black radiation absorbing substance which heats water or any other medium used for the heat transfer.

High temperatures require lenses or reflecting mirrors which capture direct solar radiation and must have tracking equipment to keep them facing the sun. Low temperatures can be used for water and space heating or for sea water distillation; high temperatures are needed for driving engines or pumps or for direct conversion into electricity by thermoelectric generators without passing through a mechanical energy stage. Solar radiation arriving in the form of light may be converted directly to electricity by means of photoelectric cells; it is most prominently utilized by nature in photosynthesis.

The discussion in this book is limited to low temperature solar thermal processes. Occasional note is made of solar thermal conversion and photovoltaic processes where pertinent. Past, present and proposed experiments are reviewed and several feasibility studies are discussed.

A section on available solar hardware has been included. While a great deal of machinery is being developed by the various companies participating in the research described, only a limited amount of equipment is commercially available. For the reader who would seek further information, a list of names and addresses of persons and companies closely associated with solar-heating-and-cooling-for-buildings research is also included.

Solar energy can be utilized for heating and cooling of buildings by placing flat plate collectors either on the roof or the side of a building (the south side in the Northern Hemisphere, the north side in the Southern Hemisphere). These collectors are fairly simple in construction. A black surface is used to absorb the sunlight; this surface is covered with one or several panes of glass which reduce radiation.

The collector is insulated on the sides and back to prevent conduction and convection losses. Water, air or some fluid is passed through the collector and can reach temperatures from $140°F$ to greater than $200°F$. The thermal energy from the fluid is then stored in a heat storage container to provide energy for the day/night cycle. The thermal storage can be sensible heat of water or rock or the latent heat of fusion of special salts.

Coupled to the heat storage system are a heating loop and a cooling loop. The heating loop takes heat from thermal storage to operate an absorption or mechanical air conditioning system. Also connected to the heat storage loop is an auxiliary heater; in most parts of the country a conventional heating system will be necessary as an auxiliary source. This makes first costs for heating and cooling more expensive than those to which the consumer is accustomed, but the over-all costs, of course, will be markedly lower.

To date most of the research done has been on residential solar-heating systems with only a small effort being expended on solar air conditioning. As major companies become involved in the feasibility studies, critical design factors will be identified and systems evaluations will be made which will no doubt improve solar technology, since the year-round use of any climate control system tends to make it more economical.

The availability of computers makes the use of simulations and models an attractive tool for the optimization of performance and the minimization of costs. Solar heating components lend themselves to these studies readily and thus will provide necessary statistical data as the next logical step to follow the early experiments, some of which were not fully documented. Large scale applications on schools and other buildings are beginning to be studied or are in the planning stages. Though considerable work remains to be done, the field and the market should be expanding rapidly as the public becomes more informed of the potential, economy and utility of solar energy for heating and cooling of buildings.

Components
for Solar Heating and Cooling Systems

BACKGROUND

The heating of buildings in the United States requires one-fourth of all the
fuel consumed in this country. Maintenance of winter comfort thus requires
the annual consumption of coal, oil, and gas having a heating value of roughly
10,000 trillion Btu. This great quantity of heat is actually used at very modest temperatures, only 21° to 32°C. In a sense, this is a wasteful use of these
concentrated energy sources which are capable of delivering heat at far higher
temperatures.

Simple equipment can be employed in the capture of solar energy at temperatures well above these requirements, so the large quantity of energy used in
space heating makes this application of solar energy highly attractive. Only a
small fraction of total electric generation is employed for air conditioning. In
the early 1960s residential air conditioning required something less than 1%
of total electric output. However, this use is growing very rapidly, and electrical demand resulting from domestic air conditioning is approaching high levels in some regions. Here also the possibility of solar-operated cooling units
suggests itself both from the standpoint of an attractively large application and
also from the energy-conservation point of view (1).

Just over 30 years ago the first solar-heated laboratory was constructed, and
the first solar-heated dwelling was put into operation about 25 years ago.
Since that time, a number of groups throughout the world have undertaken experimental work in this field, most of which has led to the design and construction of buildings heated by solar energy.

These earlier efforts may be summarized, however, by observing that practically all of them showed that by use of several variations of the flat-plate solar

5

collector, in combination with various types of thermal storage materials, a portion of the heat requirements of small buildings in the temperate zone could be conveniently supplied by solar energy. Also indicated in nearly all of the studies were the needs for short term thermal storage, auxiliary heat supply, and a reduction in the initial cost of the solar heating system (2).

There are several similarities between devices used in water and space heating and the two may in fact easily be combined. Basically, the system consists of circulating water or air through a black flat-plate collector in order to remove the heat, which is then carried into the house or into a storage tank containing water, crushed rocks or chemicals capable of absorbing heat and later released for useful purposes. The system may in some cases be operated in reverse in the summertime so that warm air or water is drawn through the collector or other part of the roof for moderate cooling through night radiation. The various systems in operation differ in details.

Any improvement in the present methods of storing solar energy will have wide applications and open the way to extensive and more economical use of the sun's heat. Space heating is one of the most important fields to offer the simplest direct use of solar energy, since only a relatively small increase in temperature is needed. Hence considerable attention has been given during the last several years to the importance of storing the sun's heat for later use in supplying vital heat when it is not so available. The storage problem for some localities may be for short periods only, i.e., storage during daytime and use during the night.

The designs tried so far include a flat-plate collector, a storage system, and a means of conveying heat from collector to storage. For reasons of economy, the collector is usually designed to act as the roof of the building, which brings up the problem of proper architectural design regarding favorable angle of roof tilt and the correct orientation. The major problem in the design of a storage system is the selection of material in which the heat energy is to be stored, since it determines the capacity of the system.

The materials tried can be divided into two broad types: those that store energy in the form of sensible heat, and those that undergo a change of state or physicochemical change at some temperature within the practical range of temperatures provided by the solar heat collector, varying between $90°$ and $120°F$.

The principal components of a solar heating system for a building are closely analogous to those of a conventional system, with one additional component. In conventional design, there is the energy-conversion or heat-transfer unit commonly known as the furnace, the function of which is the transfer of chemical energy in the fuel to a medium such as air, water, or steam. In most systems, there is also a pump or blower to circulate the heated fluid to and from the space to be heated. Usually there are pipes or ducts through which the heated fluid or air is circulated and registers or radiators from which heat is transferred to the rooms. Thermostats and other control elements regulate the supply of fuel and the movement of the heat-transfer medium as required.

The basic components required for solar heating and cooling systems are described in PB-235 428, *Solar Heating and Cooling of Buildings, Phase O, Final Report,* Volume II (Appendices A-N) prepared by Westinghouse Electric Corporation for the National Science Foundation, issued May 1974 and AD-778 846, *Solar Energy,* by V.A. Stevovich of Informatics, Inc. prepared for the Air Force Office of Scientific Research Advanced Research Projects Agency, issued in March 1974. The material in this chapter has been derived from both of these reports.

COLLECTORS

In a solar heating system, one of several types of solar collectors serves the purpose of a heat exchanger in which solar energy is employed for heating the transport medium, usually air or water. The solar heating system may then supply heated fluids or air through pipes and ducts being moved by fans, pumps, or other equipment.

In general, solar radiation received on the earth's surface is not hot enough for technical use and must be concentrated by focusing from a large area onto a small heating surface, or must be retained in a heat trap for longer preservation. The focusing type collectors provide the higher temperatures needed for operating engines, but require direct sunlight and a sun tracking mechanism.

Focusing collectors have mirrors arranged usually in a parabolic form, to focus onto a small boiler or heater. Nonfocusing collectors do not provide high temperatures and intermediate results are obtained by a permanently fixed flat reflecting surface tilted at an optimum angle equal to the latitude on the earth's surface.

These collectors usually have two or more layers of glass plates, which allow the sun's radiation to pass through but are opaque in the infrared and form a heat trap which keeps the heat-absorbing receiver from cooling off too rapidly, either by radiation in the infrared or by wind currents and convection (3).

There are three basic categories of collectors: flat plate, focusing (or concentrating), and photovoltaic. Only flat plate collector concepts will be discussed in this section for the following reasons. While many different flat plate collectors have been built and used successfully in water and space heating applications, few or no focusing collectors have. At this time, it is believed that focusing collectors should not be considered because:

(1)   They are not ready for low-cost quantity manufacture.
(2)   In use, they must be continuously movable to follow the sun within about one degree of angle.
(3)   Their design and application is too complex for one to

expect high reliability and low maintenance at low cost.

(4)   While their higher temperature output improves air conditioning operating efficiency, it is not required for adequate operation.

(5)   High temperature and pressure operation adds a substantial danger factor to housing application. Installations
would probably require constant supervision and maintenance.

Because their high temperature output is advantageous in operating air-conditioning equipment, they may, however, be economical for large buildings, which
often have a net cooling requirement most of the year. Photovoltaics can serve
the dual purpose of delivering heat and electricity, but they are presently too
costly to be considered for widespread use.

Large scale photovoltaic solar energy conversion will become an economically
competitive alternative to fossil fuel and nuclear electrical power generation
when a stable, long-lived, inexpensive solar cell having an efficiency in excess
of 5% becomes available. The problem with available silicon cells is that,
while they are stable, long-lived, and have a good efficiency (greater than 12%
AM1), they are presently too expensive. The problem with the thin-film
Cu-CdS cell is that, while it promises to be inexpensive when manufactured in
large quantities, its output is not stable over a period of decades and its efficiency is only around 5%.

General Description

The most common flat plate collector consists of a metal (copper) plate which
is painted black on the side facing the sun and thermally insulated on the
edges and on the reverse side (glass wool). Above the absorbing plate, spaced
an inch or so apart, are one or more glass or plastic surfaces to reduce upward
heat losses. The collected energy is removed by circulating water or some
other working fluid in tubes which are in thermal contact with the absorber
plate, or by circulating air past the absorber.

Schematics of several flat plate collectors are demonstrated in Figure 1.1.
Figure 1.1b is the conventional type described above; Figure 1.1a uses air instead of water for heat transfer; Figure 1.1c uses the glass shingle collector
where heat is absorbed on the blackened bottom third of tilted glass plates
and transferred to the air drawn down through these shingles and in Figure 1.1d
solar energy is absorbed on black gauze.

In general, the flat plate collector is a simple, rugged device which, without
orientation mechanisms, can efficiently collect solar energy at moderate temperature levels. It has played an important role in the history of solar energy
utilization and holds a position of great significance today. Flat plate collectors are economic media in connection with space heating and air conditioning,
and supplying hot water for domestic use.

FIGURE 1.1: SCHEMATICS OF VARIOUS FLAT PLATE COLLECTORS

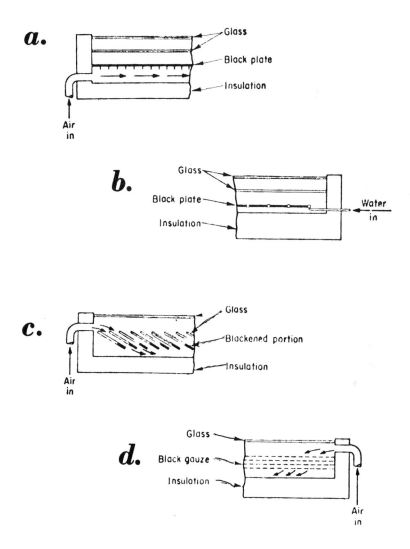

Source: AD-778 846, March 1974

They have also been used with systems for the conversion of solar energy into
mechanical and electrical energy. Flat plate collectors, unlike concentrating
collectors, can take advantage of the diffuse component of scattered solar radiation as well as the direct component. The power absorbed per unit collector

area for either one of these components is simply the product of the coefficient of absorption for solar radiation, the effective transmittance of the cover plates, and the solar incident radiation, falling on the tilted collector surface.

The heat loss to the environment is made up of conduction loss from the back of the absorber plate through the insulating material (losses through the edges of a well-designed collector are negligible) and an upward radiation conduction-convection loss through the cover plates. The per unit rate of energy loss through the insulation is dependent upon the thermal conductivity and the thickness of the insulation as well as the difference between the arithmetic mean absorber temperature and the temperature of the back of the frame. In general the cover plates are, for practical purposes, opaque to the long wavelength thermal radiation corresponding to the operating temperatures of the covers and the absorber surface (4).

The basic flat plate collector (FPC) discussed below consists of

> an absorber, heat exchanger surface which may have imbedded or bonded tubes for heat exchange to water, or it may have plates, fins, or be a porous mat for heat exchange to air;
>
> glazing surfaces of glass or transparent plastic to control upward heat loss from the absorber;
>
> insulation to reduce back and side losses; and
>
> a containing structure which may also be part of the structure of the building.

Figure 1.2 shows a basic general block diagram of energy flow and component parts in a flat plate collector. Most FPCs which have been studied and/or tested by workers in the field will fit this diagram. However, component parts and energy flows cannot be separated as shown for some collector configurations. This is particularly true for the matrix-absorber heat exchanger portion of most collectors which will be called the receiver. A discussion of energy capture mechanisms and factors influencing losses is given in the following sections.

The primary objective of the collector is to convert solar radiation into useful heat. Following Whillier (5), the net rate of useful heat delivered by the collector, q_u, is defined by the heat balance equation,

$$(1) \qquad q_u = q_a - q_l$$

where q_a is the total radiant energy absorbed by the collector, q_l is the total heat energy lost by the collector, and all terms are taken to be normalized for unit of collector area. This expression may be rewritten as

$$(2) \qquad q_u = fl - U_L (T_c - T_a)$$

FIGURE 1.2: PRINCIPLE ENERGY FLOWS IN FLAT PLATE COLLECTORS

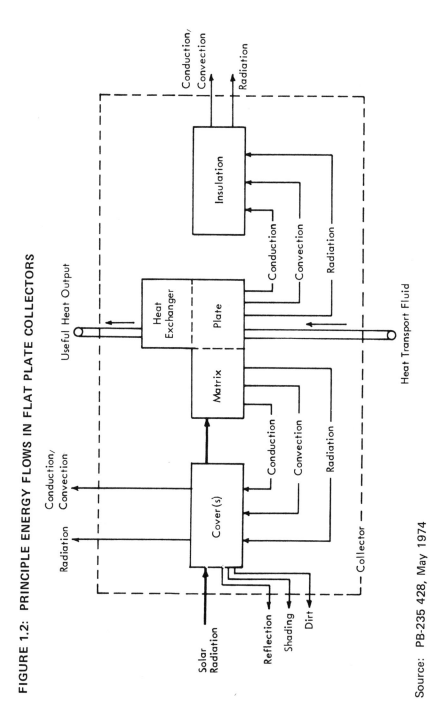

Source: PB-235 428, May 1974

when I is the incident solar energy density, f is the fraction of incident solar energy absorbed in the collector, U_L is a loss coefficient representing losses between the receiver at average temperature T_c, and the ambient air at temperature T_a.

Clearly, one would like to simultaneously maximize f and minimize U_L. The factor U_L is a function (although sometimes a weak one) of the temperatures T_c and T_a and other ambient conditions, so the equation is not strictly linear. Collector heat energy output increases as the operating temperature decreases; however, high system temperatures may be desirable since they result in more efficient operation of air conditioning equipment and a reduction of thermal storage size.

Factors influencing f are discussed in the section on optical losses and factors influencing U_L are discussed in the sections on radiation and conduction/convection losses below.

Loss Control Concepts

Losses may be divided into three categories: optical, radiation, and conduction-convection. Conceptual methods of controlling these losses are listed and discussed in the following sections.

Optical Losses: One or more of the four concepts discussed below may be used to reduce optical losses. Although factors such as shading and dirt might be considered as optical losses, their effect is not large (5). Optical losses are defined here as those which reduce the amount of heat being absorbed by the collector, i.e., they are "off the top" losses, as opposed to the heat loss after absorption by the receiver.

The optical loss may be defined as (1 – f) where f is defined above. By considering Equation 2 above, one sees that if f is decreased by, say 5%, q_u will be reduced by more, possibly 10%, after the subtraction of the heat loss term. However, a 5% increase in the heat loss coefficient, U_L, may reflect only a 2 or 3% reduction in q_u.

Low Reflectivity Covers — The reflectivity of the outside glazing plays an important role in the performance of solar collectors. At normal incidence, glass reflects about 4.3% (5,6) of the incident solar radiation at each glass-to-air interface, or about 9% per glass plate. One to three percent more is absorbed (turned into heat) in the glass.

Reflection losses are about doubled for radiation incident $60°$ from normal. Glass especially treated with a thin film or by light etching, may reflect only two or three percent (6) of the incident radiation. Performance of a collector with a single glass cover may not be very sensitive to this change in reflectivity of the glass; however, a collector with three or four glass cover plates can lose a significant portion of incident solar energy just by reflection.

High Transmissivity Covers — The transmissivity of glass or plastic covers can have a pronounced effect on collector performance. Green-tinted glass may have an extinction coefficient of 0.8 per inch, which means that 10% of the incident energy (at normal incidence) will be absorbed in $\frac{1}{8}$ inch of glass. However, clear (white) glass having an extinction coefficient of less than 0.2 per inch is easily obtained, which means that only 2.5% (normal incident radiation) is absorbed in the glass.

This may double at incidence angles of $60°$, and four cover plates could quadruple the total figure. One opposing consideration is that radiation absorbed in the cover plates may not be completely lost to the collector as it serves to raise the temperature of the cover plates and thereby reduce the outward losses. It is the above considerations which may lead to, for example, a two-cover plate collector being more suitable in a given climate than a three-cover plate collector, even when costs are ignored.

High Absorptivity Plate — As used here, the absorptivity of the collecting surface is the weighted average of energy absorbed at all wavelengths of the incident solar radiation. For an opaque surface, the absorptivity, a, is related to reflectivity, r, by

$$a = 1 - r$$

The net absorbed energy is somewhat increased by re-reflection at the glass plates. Therefore, the fraction of normally incident radiation absorbed in the receiver may be written (5).

$$F_c = 1.008 \, T a \qquad \text{(1 cover)}$$
$$= 1.012 \, T_1 T_2 a \quad \text{(2 covers)}$$
$$\text{etc.,}$$

where the constants in front, slightly greater than 1.0, account for multiple reflections, T_i is the effective transmittance of the ith cover plate and accounts for both transmissivity and reflectivity, and a is the effective absorptivity of the receiver.

This fraction, F_c, may be further modified to account for heat absorbed in the cover(s) to result in an effective transmissivity absorptivity product. However, the point to note is that the received energy is nearly proportional to τa, where τ is the effective transmissivity and a the receiver absorptivity. Therefore, both τ and a should be made close to unity and efforts should not be directed, for example, to increasing a beyond 0.95 when τ might be only 0.8.

In terms of state-of-the-art, a flat black surface finish may easily have an absorptivity of 0.95, so high values of a are not unreasonable. However, the surface reradiation, discussed in the next section, is related to a, and any attempt to reduce infrared emissivity leads to somewhat reduced absorptivity. There are instances, then, when a may be as low as 0.85, but collector performance still may be very good if emissivity is 0.1 or less. Collector absorptivity may be, like the glass transmissivity, dependent on angle of incidence, but the mathematics quickly become very complex when effects of angle are included in all factors. Therefore, reflectance, transmittance, and absorbance are usually specified for normal incidence and corrections are applied to account for other angles of incidence.

Radiation Heat Losses: Methods used to control radiation losses may be classified as spectrally selective surfaces, radiation trapping surfaces, and radiation shielding.

Spectrally Selective Surfaces — Selective surfaces have been widely studied and reported in the literature (6, 7, 8, 9, 10, 11, 12, 13). Tabor (11) provides a good discussion of the basic principles involved. For any material and for monochromatic radiation, the sum of absorptivity, a, (or emissivity, ϵ), transmissivity, τ, and reflectivity, r, is unity. Additionally, these three properties are a function of wavelength, λ, thus,

$$a(\lambda) + \tau(\lambda) + r(\lambda) = 1$$

and

$$a(\lambda) = \epsilon(\lambda)$$

Some materials, such as an absorbing plate, are perfectly opaque ($\tau = 0$) so only a (or ϵ or r) need be defined over all wavelengths of interest. A selective surface is one for which absorptivity (and emissivity) differs with respect to the wavelengths of solar radiation (shorter than 2 or 3 microns) and the wavelengths of infrared radiation (longer than 2 or 3 microns).

Radiative loss is described by the emissivity of the surface which is its radiation relative to that of an ideal black body at the same temperature. Thus, a surface may have an (infrared) emissivity much lower than its (solar) absorptivity, but as stated above, reducing ϵ by surface treatment somewhat reduces a also. Four methods of producing selective surfaces have been reported in the literature. They are:

> Thin Films — A high-reflectivity metallic surface is covered by a thin film of material opaque to visible light. The film thickness is on the order of one visible light wavelength (0.5 microns). The thickness is sufficient to absorb visible light efficiently, but long wavelength thermal radiation is reflected by the base and therefore radiated very inefficiently because the film thickness is a small fraction of an infrared wavelength.

> Geometric Trapping — A metallic surface is polished on the scale of 5 micron wavelengths but pitted or scratched on the scale of 0.5 micron, so it absorbs visible light. Alternately, a base is covered by a layer of finely divided metal such as gold, where the particle size is such that, for visible light, the layer is virtually black but at longer

wavelengths the true reflecting character of the metal is dominant.

Interference Filter — By vacuum deposition of several layers of suitable thicknesses (on the order of ½ wavelength) and with different indices of refraction, it is possible to achieve high reflectance in the infrared and low reflectance in the visible. By using more layers, say 4 or 5, the transition can be made to occur more abruptly and at any desired wavelength.

Semiconductors — A semiconductor, which is intrinsically opaque to photons having energies higher than the band gap energy but transparent to photons having energies lower than the band gap energy, is deposited on a polished metal base. The semiconductor opacity dominates at visible wavelengths (higher energies), whereas the polished base reflectivity dominates at infrared wavelengths (lower energies).

Characteristically, selective surfaces exhibit certain shortcomings. As the emissivity, ϵ, is reduced, so is a. Special precautions must be taken to avoid deterioration of the optical and/or thermal properties of the surface due to oxidation, chemical changes, etc. Selectivity is often strongly angle dependent, so absorptivity is lower at other than normal incidence.

A carbon black surface treatment could be expected to have $a = 0.95$, $\epsilon = 0.95$. Selective surfaces which until now have been made only in small quantities have $a \approx 0.85$, $\epsilon \approx 0.1$. The selected surface manufactured by Alcoa is estimated to have $a \approx 0.9$, $\epsilon \approx 0.2$. NASA has reported a surface with $a > 0.9$, $\epsilon < 0.05$ and Honeywell has in use a nickel black surface with $a \approx 0.94$, $\epsilon \approx 0.07$.

Radiation Trapping Surfaces: Some researchers have studied special collector geometries designed to enhance effective absorptivity and/or reduce effective emissivity. Perhaps the most intensively studied concept has been the use of "honeycombs" or cellular structures. Hollands (14) studied the directional properties of vee-corrugated surfaces. Chiou et al (15) studied a slit and expanded aluminum foil matrix, and Swartman and Ogunlade (16) studied packed-bed (spheres, screens) collectors. Bevil and Brandt (17) investigated parallel fins of aluminum, and other workers (18, 19, 20) have studied porous bed collectors.

Some of these configurations have been selected for reasons more compelling than simple radiation loss reduction; however, they are included here because they do fit more or less into the category of radiation trapping surfaces.

Vee corrugations increase absorptivity geometrically by requiring the sun's rays to undergo two or more reflections; however, this also increases the effective emissivity of the surface. For example, a $30°$ Vee, made from a surface having $a = 0.8$ and $\epsilon = 0.05$ would result in an effective absorptivity of 0.99 and a hemispherical emittance of 0.18 (12).

The other radiation trapping surfaces, mentioned previously, are more qualitatively selective than quantitatively selective, since few absorptivity values appear in the literature. For example, a porous bed will admit some sun rays to depth in the bed. If air flow is in a direction so that the coolest (entering) air is exposed to that part of the bed facing the sun, then the exposed "effective surface" of the bed will be cooler, while the geometrically hidden part of the bed will be hotter. The effective cool surface results in a reduction of radiation losses even though effective values of a and ϵ are difficult to calculate or measure.

Trapping geometries such as honeycombs, fins, and vee-corrugations, have the disadvantage of being very much poorer traps when the sun is a few degrees away from normal incidence. The angle of intercept is determined by the geometry, but some of these designs are unsuitable for stationary mounting.

Radiation Shielding — The principal method of radiation shielding is the use of multiple cover plates. Between a hot surface and cool surface there is an effective heat transfer resistance which depends on the emissivities and temperatures of the two surfaces. Multiple covers essentially stack these resistances in series, thus reducing radiation loss from the absorber to the sky. It should be noted that, while glass is nearly opaque to thermal radiation, many plastics (especially films) are not. Some plastics, therefore, may be more important in controlling convection losses than in controlling radiation losses.

Conduction/Convection Losses: When radiation losses from a collector are reduced, convection quickly becomes the dominant loss. Convection losses control unless the plate-to-plate spacings are reduced to one-half inch or less; then conduction loss dominates. The collector should be designed to prevent significant quantities of heat from reaching the outside cover since that heat will be quickly conducted through the cover and convected into the ambient air. The following concepts of conduction/convection control can be employed.

Multiple Glazings — Multiple glazings can be closely spaced (on the order of 1 cm) to provide a stagnant air gap (5, 21). When the air is stagnant, convection across the gap ceases and only conduction remains. There is a trade-off, of course, in the width of the gap. A small gap will stand a higher gradient before convection sets in, but conduction losses are inversely proportional to the width of the gap. However, while convection loss calculations for horizontal air gaps heated from below are quite valid, considerations similar to those of Charters and Peterson (22) may show that a stagnant gap in a horizontal collector may not be quite so stagnant when the same collector is tested.

Between parallel plates, the combined conduction/convection loss becomes very nearly independent of plate spacing when the separation is greater than 2 or 3 cm. Thus, unless convection exchange can be eliminated, the loss is near minimum for 2 or 3 centimeter spacing with the thermal gradients encountered in flat plate collectors.

Honeycombs or Cells —The honeycombs discussed above are useful in the control of convection losses as well as the control of radiation losses. Cells have been investigated by many authors (8, 19, 22, 23, 24, 25, 26, 27). Tabor (27) points out that the characteristic length used in computing Grashof (or Rayleigh) numbers for convection loss computations is the cell's diameter. For ordinary flat plate collectors, convection is suppressed for Grashof numbers less than 2,000 which, in practice, means air gaps of about 1.25 cm or less.

For temperature differences on the order of $50°C$, experiments have shown that a 1 cm diameter cell just suppresses convection (27, 28). Therefore, the cells can be made narrow (to prevent convection) and long (to reduce conduction). However, Charters (22, 23) has shown that while vertical cell honeycombs in horizontal collectors are quite effective, the same cells in tilted collectors can be quite ineffective.

Evacuation — At pressures in the range of 10^{-2} to 10^{-4} atmospheres, convection is suppressed and the heat transfer is determined by the conduction of the air. However, the thermal conductivity of the air is not reduced until the mean free path of molecules becomes comparable to wall spacing. As a result, for vacuums of 10^{-5} to 10^{-6} atmospheres, the thermal conductivity is reduced to about 10% and for another decade reduction in pressure (10^{-4} mm of Hg), the conductivity is about 1% of that of air at atmospheric pressure.

Evacuation is certainly a state-of-the-art technology and the principal factor in design is to provide long lasting glass to metal seals. This seal may be destroyed by differential expansion between the metal and glass.

Speyer (29) designed and built metal collector surfaces contained in long evacuated glass tubes about 2 inches in diameter similar to fluorescent light tubes. The absorbers in the glass tubes were made from steel tubing pressed together in the center to form an elongated "U" for water passage. Speyer showed some alternate conduit designs, used a selective black, and experimented with silvering on the lower half of the tube to augment radiation and reduce radiation losses. Other refinements are perhaps possible.

The important point in evacuated collector design is that a highly selective surface must also be employed. If the emissivity is not less than 0.4 (Speyer's surface) the performance may be no better than a collector with two or three glass plates and no evacuation.

Counterflow Exchange — The concept of counterflow exchange is employed to reduce convection losses in some air heating collectors. The idea is to force incoming (cooler) air in a downward or transverse direction so that heated air does not reach the collector cover. In a sense, the heated air may preheat the incoming air. In any event, the collectors can be designed so that the air flow reduces the upward convective heat transfer.

Some examples of air collectors employing this method are the two-pass collector (30), and the porous bed air collectors (5, 18, 19, 20). The overlapped glass plate collector (31, 32) has some aspects of this design, but it also relies on laminar flow between plates to reduce upward heat transfer.

Internal Heat Transfer Concepts

Since the function of a collector is to heat a fluid which is flowing through it, it is important to design for minimum temperature drops between the absorber and the fluid. Methods of maximizing the useful heat q_u were discussed above and an average receiver temperature T_c was defined. This section is concerned with concepts of transferring that useful heat to the fluid entering the collector at temperature T_{fi} and leaving the collector at temperature T_{fo} (having a temperature rise of ΔT_f).

Water Heating Collectors: Most water heating collectors have a variation of a "tube in plate" absorbing surface. The plate absorbs solar energy and conducts the heat to water contained in tubes or internal passages. It is easy to show that a uniformly heated metallic tube will conduct the heat into the water with little temperature drop (maybe 2° or $3^\circ C$), so once the tube is heated, the heat transfer to the water does not present a significant design problem.

The design problem which soon becomes apparent is the transport of heat transversely across the plate to the tube and, in some cases, through the bond between the tube and the plate. The economic trade-offs are thickness of plate and spacing of tubes. A thicker plate will conduct heat better but is more expensive, and closely spaced tubes will allow a thinner plate but will require more tubes per unit area.

The performance equations for several tube-in-plate configurations provided by Whillier (5) include fin efficiency (transverse transport of heat in the plate) and bond conductance (where the tubes attach to the plate). Holman (33) provides a background for analysis of other configurations. Flow through the tubes is almost always slow enough to be laminar, and calculations pose no great problems.

In terms of overall collector performance, the designer needs to be aware that, while losses do not outwardly appear in these heat transfer calculations, a higher plate-to-fluid temperature difference necessarily implies lower collector efficiency, since radiation and conduction/convection losses increase with increasing average plate temperature. In contrast, system efficiency, in general, increases with higher fluid temperatures, which necessarily implies a higher average plate temperature.

Therefore, the collector must be designed to minimize plate-to-fluid temperature differences, and the system must be designed to optimize overall performance in terms of collector operating temperature. These considerations do not produce (with the exception of transverse fin efficiency heat flow) serious problems in water heating collectors, and the average operating temperature of the collector can be simply adjusted by varying fluid flow rate.

Air Heating Collectors: Air heating collectors can have significant temperature differences between average collector temperature and average fluid temperature. Air heating collectors are subject to an additional trade-off. If it is desired to increase the air exit temperature, the air flow rate should be reduced; however, as the flow rate is reduced, the air flow becomes less turbulent (or even laminar) and the heat transfer coefficient between the air and the collector is reduced. There is thus a compounding effect of flow rate adjustment.

Three conceptual methods can be employed which increase the transfer of heat between the receiver and the air: increased heat transfer surface area, increased air turbulence, and multiple pass air flow (a type of increased surface area). The most important of the three is, of course, increased surface area. Workers have investigated porous beds or matrices (5, 15, 18, 19), fins (17), multiple passes (30, 34), vee surfaces (30, 35), and roughened surfaces (36). Most of these concepts also rely on air turbulence to increase heat transfer. The overlapped glass plate collectors (31, 32) are designed for laminar, but two-sided, flow across the absorbing surfaces; laminar flow was selected to reduce upward convection losses.

Special Problems

Several problems, not directly affecting calculated performance, are listed and discussed below. While most of these problems are not so important when considering performance concepts, their effects on final collector and system designs are very significant.

Freezing and Boiling in Water Heater Collectors: A collector can be drained at night to prevent freezing, but this aggravates corrosion problems, and sometimes causes problems in completely and automatically refilling all tubes to assure even flow. The fluid can be charged with an antifreeze (ethylene glycol is most commonly mentioned) solution, but this requires a separate heat exchange loop to storage if the system is using water storage. It has been suggested that slow circulation from storage could be used to prevent freezing, but it has not yet been shown what effect on system performance this might have (additionally, a pump failure could then have disastrous effects on the collector).

Boiling causes problems in a tightly sealed system, yet it is desirable to have a sealed system from the corrosion standpoint. One may have to design the system to withstand the pressures attending the highest temperature the collector can reach. For $300°F$ this is about 59 psia, with no safety factor. Over-temperature relief systems such as steam release, increased convection cooling, shading, etc., should probably be designed into a collector because designing and building the collector to hold the required pressure could cause safety hazards.

Corrosion: Copper tubing may be used to prevent corrosion in a water collector, but copper is expensive. Aluminum and steel are the next choices, and both of these materials are subject to corrosion. Suitable inhibitors may be used to prevent corrosion, but the inhibitors are somewhat less satisfactory for aluminum. The corrosion problem is aggravated by draining, since the metals are then exposed to oxygen in the air.

Damage by Heating: Materials in contact with the absorber surface need to withstand temperatures (perhaps 200°C) which might be encountered as a consequence of lost fluid circulation. High temperatures can also cause degradation of selective absorbing surfaces.

Damage by Hail: Since nearly all collectors considered to date (and likely to be considered in the near term) have glass covers, glass breakage could be an important consideration in some locations. Double strength glass can be used to resist breakage, but this glass is, of course, more expensive. Even so, it is important to construct the collector so that the glass can be easily replaced, and so it is well supported and not in too large sections. Architectural standards would apply here.

Contamination in Air Collectors: Dust, moisture, and pollutants will circulate through an air heating collector and can conceivably cause surface deterioration, clogging of small passages, etc. This becomes serious if an air heating collector makes use of a selective surface, since selective surfaces are often delicate to begin with.

Deterioration of Paints and Other Materials: Since the collector in a system will involve a high capital investment, it is desirable to design for at least 20 year life. The relatively high temperature environment inside the collector poses special problems in materials. A black paint (which might be used to avoid instabilities of a selective surface) might become faded, cracked, etc. in the long term. Therefore, special materials must be used, or simple maintenance procedures must be provided.

Heat Transfer Fluids: Air and water have been mentioned as the most suitable heat transfer fluids. Water must be protected from freezing, and as a result the heat carrying capacity is reduced 10 to 20%. Other liquids could be considered for heat transport from the collector. Any of them would have a lower specific heat than water, often half or less, but some have higher boiling points (lower vapor pressures) and may also have lower freezing temperature. With liquids other than water, the freezing is not likely to damage the collector, anyway, because they shrink when freezing.

Materials Used for Flat Plate Construction

Some of the important materials which have been used in flat plate collectors are discussed below. The types of application, the advantages, and some limitations to the use of each of the materials are discussed.

Aluminum: Aluminum appears to be suitable for use in all subsystems in the solar heating and cooling of buildings. Aluminum sheet, plate, tubing, or pipe can be used in solar collectors. Aluminum sheet, plate, tubing, or pipe can be used in solar collectors. Aluminum can be given many types of surface finishes, including a selective black, and it is an excellent base for paints, if organic type finishes are used. Aluminum can also be used as the heat storage containers for hydrated salts, water, or stones.

Its high thermal conductivity could be advantageous for containers of hydrated salts, but corrosion, as discussed below, could be a problem. With respect to corrosion, some problems also could occur with certain waters, and this is discussed below. Aluminum piping or ductwork could be used to transfer the heat from one component to another. Aluminum heat exchangers currently are used so they would appear applicable to solar heating and cooling systems. Overall, fewer corrosion problems probably would be encountered with aluminum or other metals if air instead of water served as the heat transport medium.

For auxiliary roles, aluminum could be used as structural or framing members for solar collectors because it has a high resistance to atmospheric corrosion. If required for aesthetic reasons, such members could be given any one of a variety of finishes. Aluminum foil has in the past been used for thermal insulation, and it could be considered in this application.

Corrosion Resistance — Water: Aluminum has been used for years to store and transport high purity (distilled or deionized) water. Many potable waters have no effect on aluminum, but some do. The most common cause of corrosion of aluminum by potable waters is from naturally occurring or contaminating heavy metals such as copper, lead, and nickel. These are usually found in acidic waters and cause scattered localized attack of the aluminum. Less frequently, corrosion of another type (etching) results from water with an alkaline pH greater than 9. The nature of the compounds causing the alkaline pH determine whether or not attack occurs.

To combat localized corrosion from potable waters, an alclad aluminum is recommended. The cladding alloy is anodic to the core alloy, and this means that corrosion proceeds laterally along the alclad surface instead of penetrating into the core aluminum alloy. If the waters are recirculated, corrosion inhibitors such as chromates or soluble oils can be added to waters containing heavy metals, or silicates can be added to corrosive, alkaline waters. Changing the pH of recirculated water by suitable additives can also be beneficial. Adding inhibitors or changing the pH of once-through waters generally is not economically feasible.

Ammonia: Aluminum alloys are highly resistant to corrosion by aqueous solutions containing 60% or more ammonia. Aluminum alloys 3003 and 1100 have been used for many years for refrigeration coils with anhydrous or nearly anhydrous liquid ammonia refrigerant. Aqueous solutions containing 30 to 60% ammonia have very slight action on aluminum, with the main problem being small amounts of hydrogen generated by the corrosion. Aqueous solutions containing less than 30% ammonia corrode aluminum, with the corrosion rate peaking at about 5 to 10% ammonia. Aluminum alloys are compatible with gaseous ammonia at temperatures up to 450°C and to hot gaseous ammonia containing air.

Salts: No information is available on the effect of $Na_2SO_4 \cdot 10H_2O$ (sodium sulfate) or $Na_2HPO_4 \cdot 12H_2O$ (sodium phosphate) on aluminum at elevated temperatures. However, both chemicals are considered noncorrosive when they are placed in small aluminum cups stored one month at approximately 100% relative humidity at room temperature.

No tests have been conducted with lithium bromide solutions, but tests with solutions of related chemicals have indicated that they mildly pit aluminum. Sodium chromate additive inhibits the corrosion, so a similar addition to lithium bromide solutions probably would have the same inhibitive effect.

Possible Limitations — With hot water, problems from the following sources should be considered, even though they are not likely to develop with suitable precautions. Galvanic corrosion from dissimilar metals in the system could be obviated by use of an all-aluminum system, by electrical separation of the metals, by painting the dissimilar metals, or by inhibitors. Scaling from water could decrease heat transfer in aluminum tubes as well as tubes of other metals. Water treatment might be required if scaling develops. Erosion could develop from water of high velocity or at constrictions or sharp bends in the tubes.

The selective black surface for aluminum developed by Alcoa does not deteriorate with time, based on the performance of selective black specimens being wrapped in paper and stored years indoors. However, these selective black surfaces freely exposed outdoors to the local industrial atmosphere degrade with some loss of absorptivity in several weeks. No glass cover plates were used in this case. The durability of the selective surface under solar operating conditions with glass cover plates requires additional testing.

Glass: Glass is probably the most suitable material known for collector cover plates to shield the absorber against radiation and convection losses. ⅛" or DS glass transmits 85 to 88% of available solar energy. Considerable data is available relative to structural loading, wind loading, impact strength, durability, and thermal stability.

It is expected that one of the future improvements or developments with glass will be in reducing the iron content to improve the transmission of solar energy. Perhaps it will be possible to increase the solar energy transmission of ⅛" glass to 92 to 94%. This, it is believed, will contribute directly to the efficiency of the system.

Further development of improvement in the basic glass cover plate would involve either antireflective or infrared reflective surface coatings; these are currently known. Whether or not they will be a reality for solar energy applications will depend, of course, on the relative economics. It has been reported by A.D. Little, Inc. that conventional flat plate collector systems to heat and cool buildings and take care of the hot water needs can be built for $7 to $8 per square foot of collector. With strong attempts to drive the costs down to $2 to $3 per square foot, it gets very difficult to justify additional items such as glass coatings and selective absorber coatings unless significant contributions to efficiency relative to capital costs can be made.

It must be pointed out that this short treatment of the subject refers to glass used in conventional flat plate solar collectors for heating, cooling, and fulfilling the hot water needs of buildings. Concentrators, hot air systems, and vacuum collectors could alter the picture significantly in some cases.

PPG Industries has been producing for years a number of products that contain materials and technology that can be used for solar energy applications. There are heat absorbing coatings, insulation, and fiber glass structural components that have proven their utility and durability by many years of exposure to the environment. They are currently evaluating these for potential use in solar energy applications.

Plastics: Plastic materials may be useful to solar heating and cooling systems in numerous ways. In general, though, the applications should not subject the material to temperatures exceeding $100°$ to $125°C$ or there may be permanent damage in the form of oxidation, change in shape, or rupture.

In collectors, plastics may be used in glazing either as films or hard sheets. Structural and support members and insulation may be plastic, and even the heat absorbing surface may be plastic tubing or mattress type bags.

Methyl methacrylate is a plastic glazing material with the highest resistance to ultraviolet degradation. Stabilizers can be added to improve its resistance, but still, its transparency may be reduced 10 to 20% in 20 years or less, and the first cost is higher than for glass. Its advantage would be in its resistance to breakage, though it is more easily scratched than glass.

Tedlar films have been incorporated in collectors as glazing, but they do not have a proven life of more than five years, perhaps because of the operating temperatures as well as UV degradation. Plastic films are less satisfactory than glass because they transmit more of the infrared radiation from the collector surface. The films have only the cost advantage in glazing applications. Any of the plastics is protected somewhat by using it behind glass, but the life cannot be extended indefinitely.

A cross-linked polyethylene tubing filled with carbon black has been produced in small quantities by Polyset, Inc. of Manchester, Massachusetts. This tubing could be suitable (to $125°C$) for collector absorber or heat transport use. It could also be used in heat exchangers because of its potentially low cost and stability in the presence of corrosive liquids and gases. Filled plastics such as this resist weathering, but their primary shortcoming is softening and possible oxidation with higher temperature operation. Foamed plastic is a very satisfactory insulation material except when subjected to temperatures of 100°C or higher, however, foamed plastics have recently been regarded in FTC complaints as fire hazards. Glass insulation is less expensive and should be used instead of plastic foams. Even when the foams add structural rigidity to the design, they are probably not the most economical choice in this application.

Plastics are probably the most suitable materials for containment of normally corrosive materials at low temperatures. However, soft plastics should not be used in contact with hot water or other chemicals because the softeners will be extracted, leaving the plastic hard and brittle. Some synthetic rubbers may be more suitable for the containment function than plastic where moderate flexibility and even stretching of the containment vessel may be required.

HEAT STORAGE

In utilizing solar energy for space heating and cooling, it is necessary to provide a means by which thermal energy collected during periods of sunshine can be stored for use at night or on cloudy days. There are numerous factors in the design of a heat-storage system. These include the storage medium, operating temperature range, heat-storage capacity, transport material, means for transferring heat from transport medium to storage and from storage to the medium heat transport to the heated space.

Forms of Thermal Energy Storage

The best understood ways to store thermal energy are by sensible heat storage and latent heat storage. TES systems using sensible heat storage utilize the solar energy collected to raise the temperature of the storage medium without changing the phase of the material. The heat storage capacity of such systems is, therefore, determined by the specific heat and density of the material used. Latent heat TES systems utilize the solar energy collected to produce a phase change (e.g., solid to liquid) in the storage material. In this type of system, the storage capacity is determined by the heat of transformation and density of the storage material.

TES Material

Before discussing specific materials for TES applications, it is helpful to introduce the major factors that should be considered in evaluating various materials. These factors are the relation to building structure and the cost of TES unit, storage capacity/unit volume, cycling life, and material cost and availability. The storage capacity of the material determines the storage temperature and volume of the material required. Many materials exhibit changes in their physical or chemical structure after undergoing repeated heating and cooling cycles. This is especially true of the phase change materials. Therefore, cycle life is important from the standpoint of frequency of replacement of the storage material.

Sensible Heat Storage Materials: The two most widely used materials for sensible heat storage are water and rock or gravel. Both materials are low in cost and readily available in all parts of the country.

Water — At the present time, water is used in approximately 70% of the TES systems using sensible heat. By adding ethylene glycol to lower the freezing point, freezing in the collector can be avoided. However, this adds the requirement for a heat exchanger between the collector loop and the storage system, since the cost of using ethylene glycol as 10 to 15% of the storage is excessive. Of secondary importance is the fact that the ethylene glycol reduces the specific heat of the mixture. The storage capacity of one cubic meter of water is 10^3 kcal/°C, whereas a 15% ethylene glycol mixture has a storage capacity of about 0.85×10^3 kcal/°C.

Water is low in cost and readily available in all populated areas of the country. Water also has the largest specific heat of any liquid or solid and can be easily stored and pumped through heat exchangers. All of these factors point to water as an attractive material in which heat could be stored as sensible heat.

The use of crushed rock or gravel for heat storage is convenient in the combination of a solar air heater and a hot-air heating system. With a heat capacity only about one-fifth that of water, five times as much gravel needs to be employed for storage of the same quantity of heat between the same extremes of temperature. Storage of 1 million Btu of heat over a 37°C temperature range would require about 25 tons of loose rock, which would occupy a bin about 8 feet on a side.

Rock — Rock and gravel TES units with air as the working fluid were used in several demonstration solar houses because of the simplicity of installation, as well as continued reliability and low maintenance. These materials are inexpensive and readily available around the country. However, large temperature drops may occur in the system when air is used as the working fluid and these losses of available energy dictate higher operating temperatures with resulting lower efficiency at the collector. Additionally, dry rock storage is only about 30 to 40% as efficient per unit volume as water storage because the specific heat of rock is about 0.2, its specific gravity is 2.7, and spaces must be left for movement of air among the rocks.

Storage in Collector — Several systems have been built which incorporate storage and collector as a unit. These schemes have included large masses of concrete or even water in front of the absorbing surface. Other collector designs have included black paraffin to store latent heat in the collector.

Latent Heat Storage Materials: Because of their higher heat storage capacities per unit volume, a great amount of interest has been shown in latent heat storage materials. The major work in recent years has involved paraffins and eutectic mixtures of salts and salt hydrates. These materials are briefly discussed below.

Salt Hydrates — Because of its low cost and large latent heat of fusion, sodium sulfate decahydrate (Glauber's salt) has received the most attention for consideration as a TES material for solar heating.

Glauber's salt ($Na_2SO_4 \cdot 10H_2O$) and sodium phosphate dodecahydrate ($Na_2HPO_4 \cdot 12H_2O$) melt and dissolve in their own water of crystallization at $32°$ and $36°C$, respectively. The latent heat effect on these processes is 104 and 114 Btu/lb, respectively. When heated by a medium above this temperature, the salts melt and absorb these energy quantities; when surrounded with a medium cooler than about $30°$, the salts recrystallize and give up the heat of crystallization to the cooling medium.

Small containers of the salt can be stacked or otherwise packed in bins or closets through which solar-heated air is circulated, thereby storing heat for subsequent use (1). The use of materials that undergo physicochemical changes has been largely proposed for the purpose of reducing storage space. In early pioneering work on solar house heating, M. Telkes introduced the idea of using a hydrated salt, such as the Glauber's salt which on heating melts in its own water of crystallization. This appeared to be a very satisfactory solution of the storage problem until it was found that these materials can be very temperamental in their behavior.

Reports (37, 38) indicate that the problems encountered make sodium sulfate decahydrate unsuitable for thermal energy storage. The major problems were due to the fact that the sodium sulfate decahydrate melts incongruently to form a saturated solution and solid anhydrous sodium sulfate. On cycling, this two-phase mixture will not completely recombine due to the settling of the anhydrous salt. This caused the latent heat storage capacity of the material to decrease significantly with repeated cycling. Various techniques were employed to prevent settling but were unsuccessful.

Paraffins — The specific heat of various paraffins at room temperature is about half the specific heat of water. The heat of fusion for these paraffins is in the range of 35 to 45 cal/g or about 75 times their specific heats. Reported tests (39) indicate that a $C_{15}-C_{16}$ paraffin mixture was the most attractive possibility for TES application.

A problem pointed out in the report was that the phase change in paraffins is accompanied by an approximate 10% change in volume. Part of this volume change occurs because air, which is very soluble in the liquid phase, is insoluble in the solid phase. An additional consideration is the compatability of paraffins with container materials. For example, environmental stress cracking makes polyethylene and polypropylene unsuitable for use with paraffins.

Disadvantages — Some disadvantages which plague many latent heat storage systems are summarized as follows: instability of solution under cycling; tendency to supercool requiring nucleating agent; low thermal conductivity of solid phase, reducing the heat transfer; shrinkage of solid phase from containment vessel further reducing heat transfer; and high cost of containment vessels, tanks, and materials.

With continued research effort these problems may be solved but Westinghouse believes latent heat storage does not yet offer a reliable system. It would therefore be difficult to show that latent heat storage offers any cost advantage over sensible heat storage systems.

Cold Storage: The storage of cold has been considered in a number of cases to provide for cooling during periods of hot weather. Detailed studies have been made regarding the use of off-peak electric power to operate air conditioning units, and using the air cooled during these hours to lower the temperature of a TES material such as rock or water.

During the hot hours, warm room air would be cooled by transferring its heat to the cold storage material. To be able to store coolness for the night hours, a solar operated air conditioning system would have to use excess system capacity over that needed for daytime cooling to charge up the storage.

Storage Facilities

The only detailed considerations of containers for the various TES materials were reported by Whillier (40) and Speyer (41). Whillier's general conclusions apply to storage of sensible heat utilizing water or rock as the TES material. Speyer investigates the cost of TES systems, specifically fiber glass-insulated steel tanks containing water as the TES material.

The storage system should be designed so that the fluid supplied to the collector is at the lowest possible temperature. Ideally, the heated fluid from the collector should enter the storage container at a point where the storage material temperature is the same as the fluid from the collector.

Ideal realization of such a variable entry point is difficult but in practice the water returned to the top of the tank settles to a level according to its density, only partially mixing during the settling.

In the simplest system, then, the fluid is returned from the collector to storage at the point where the storage temperature is the highest. If the major dimension of the container is vertical, the desired temperature stratification is encouraged.

The storage system should be located in that space in the building which is cheap to provide. It also should be located such that heat losses from storage will go into portions of the building that require heating.

HEATING AND AIR CONDITIONING EQUIPMENT

Solar Heating Requirements

Utilization of solar heat to heat a room requires a heat exchange to air which may be accomplished in a number of ways:

(1)   Exchange the heat to air in the collector and then duct the heated air to the room, or

(2)   Exchange the heat to water or other liquid in the collector, then exchange to air by forced convection in a main plenum for distribution to rooms, free convection in a baseboard or wall unit, or radiant heat from pipes imbedded in the floor.

If the space is both heated and cooled, using the same heat exchange equipment for both purposes will have cost advantages. However, if air is used to transport heat from the collectors, the air conditioning/cooling cannot be done by the absorption equipment described below because of the necessity for two heat exchanges, plate-to-air and air-to-generator which gives up too much of the available energy. Maintaining high fluid temperature for space and water heating is important only in trying to preserve the available energy of fluid handled and reduce the heat exchanger area requirements.

Although a conventional system usually employs only one unit for circulating the heating medium, a solar heating system may require two or more pumps or blowers for the various requirements of moving fluid or air from collector to storage, storage to heated space, and from heated space to the other units in the system. Ducts or piping between the heated space and the system components may be similar to identical to those involved in conventional systems, and the ultimate medium for room heating may also be the same.

Auxiliary Energy Requirements: The choice of the type of auxiliary energy to operate the heating and cooling system when solar energy or stored energy is inadequate depends somewhat on the system being supplemented. In some cases, it is a good strategy to depend on stored fuel (oil or propane for example) when solar supplied heat is unavailable because this does not cause demand peaks on the public utility.

Where natural gas is available already in sufficient quantity to meet the demand, it can be used. Even electricity can be used, but it is inefficiently applied if it is used to furnish resistance heating. A heat pump furnishes several times more heat from the electricity, but it has an excessive initial cost if it is used only a few hours a year.

All these considerations and more go into the choice of auxiliary heating or cooling to assist the solar heating and cooling equipment. Even the point of application of the auxiliary energy is important to the system efficiency. In general, additional heat, if required, should be supplied as near the load as possible. Supplying additional heat to storage, for example, increases the system heat loss, and eventually less of the auxiliary energy arrives at the point of use.

An important component in the solar heating system is the furnace or auxiliary heating unit, although this would ordinarily be a conventional item of a standard heating system. There are several ways in which the furnace can be used to supplement solar heat. The objective is to have auxiliary heat available whenever needed, but to use the minimum amount of fuel consistent with human comfort in the building. A design involving completely separate water piping or duct-work systems from a furnace is not necessary and involves excessive installation cost.

Hence, the furnace should be arranged to heat the same fluid and distribute it through the same channels employed for solar heat. One of the simplest designs is a hot-air system in which air from the solar collector or storage can be passed through the furnace, if necessary, before distribution to the rooms. This arrangement permits maximum utilization of solar heat, fuel being used only when air temperatures from the solar system are inadequate for carrying the heating load.

In one design, air from the solar system always passes through the furnace, fuel being used only when required. In another arrangement, a bypass and automatic dampers divert part or all of the air through the furnace when needed. The same sort of combination may be provided in a hot-water heating system. These designs have the advantages of simplicity and the avoidance of auxiliary heat use except when actually required.

Such systems also respond rapidly to heating demands, particularly when atmospheric temperatures are subject to rapid change. They have the disadvantage of creating maximum demands on natural gas or electric networks simultaneously with maximum demands elsewhere in the area, and do not permit economies which might be obtainable by use of off-peak energy.

Heat Pump: The heat pump presents a number of special requirements when used in conjunction with a solar heating system. The heat pump is a refrigeration system which utilizes electric energy for operation of the compressor. When cooling is desired, it is operated just as a conventional compression-type air conditioning unit. Heat is withdrawn from the living space and rejected at higher temperatures outdoors to the atmosphere. In winter, reversal of refrigerant or air flows permits the supply of heat from the cold atmosphere to the refrigerant and the increase of this temperature by compression so that heat is then delivered to the rooms at higher temperature.

Temperature Control: Automatic control of heating and cooling systems is an essential design feature. When augmented by solar collector and storage equipment, with their fluctuating output, these systems have a complexity and variability which can be controlled only by a complete automatic assembly carefully designed for the application being made.

In addition to requiring the conventional thermostats which act to start and stop fuel supply, air circulation, hot-water circulation, and other operations in a house, a solar heating system usually must have thermostats in the solar collector and the heat-storage unit. The collector thermostat has the function of controlling the motor which drives the pump or blower that circulates the heat-collecting medium from solar collector to storage (or to the living space). This thermostat senses solar radiation and is set so that the circulating motor will be in operation whenever heated fluid can be delivered from the collector at a useful temperature.

In addition, the storage thermostat serves to actuate a motorized fuel-supply valve or fuel pump in the auxiliary furnace, as well as to control an auxiliary heater, in conjunction with a house thermostat, in supplying fuel to a furnace whenever the house, which is colder than a preset value, requires heat from storage (1).

Solar-Powered Air Conditioning

The concepts considered are absorption systems, Rankine cycle-vapor compression systems, jet ejector systems, adsorption systems, Rankine cycle-inverse Brayton cycle systems, and night radiation. There are a number of variations possible on each of these systems, but the basic mechanism of operation of each is as follows.

Absorption Systems: A solution of refrigerant and absorbent, which have a strong chemical affinity for one another, is heated in the higher pressure portion of a system (generator). This drives some of the refrigerant out of solution. The hot refrigerant is then cooled until it condenses and can be passed through an expansion valve into the low-pressure portion of the system.

The reduction in pressure through this valve facilitates the vaporization of the refrigerant which ultimately effects the heat removal from the environment. The vaporized refrigerant is next recombined with the absorbent mixture from which it was initially obtained to form a mixture which is rich in refrigerant. The mixture is pumped back into the high-pressure side of the system, where it is again heated and the cycle continues.

Although small units suitable for residential applications have not been produced, considerable operating and manufacturing experience is available on units which range in size from 15 to 1,600 tons. This experience has been obtained with lithium bromide-water and ammonia-water systems, and direct extension of the technology to 3-ton units poses no problems. Other working fluids such as lithium chloride-water, ammonia-organic solvent, sulfur dioxide-organic solvent, halogenated hydrocarbon-organic solvent have been considered in experimental units.

Rankine Cycle-Vapor Compression Systems: This system consists of the Rankine cycle heat engine which produces a mechanical output to drive a conventional air conditioning compressor. In the vapor compression portion of the system, refrigerant vapor is compressed and subsequently cooled until it condenses. This liquid, high-pressure refrigerant can then be expanded to produce the cooling effect before it is recompressed and the cycle continues. The heat engine portion of the system is connected only mechanically to the refrigeration system. It uses heat energy to vaporize a working fluid, which is subsequently expanded through a turbine or piston device to produce the mechanical power output. This working fluid is then condensed and pumped back into the vaporizer to close the heat engine cycle. The vapor compression part of this package is the most commonly used means of producing cooling today. Units are available over the range from one to 8,000 tons.

Jet Ejector Systems: A jet ejector is a device which can be used to reduce the pressure over a liquid refrigerant. This causes evaporation from and hence cooling of the liquid. Refrigerant cooled in this manner is circulated through a heat exchanger to effect the desired atmospheric cooling. Operation of the ejector is accomplished by passing high temperature and pressure refrigerant vapor from a boiler through it to a low-pressure condenser. Pumps are required to maintain proper refrigerant volumes in the vaporizer, condenser, and evaporator. Steam jet ejector systems have been employed for many years in the food and chemical processing industries to dry or concentrate foods and chemicals.

Adsorption System: The basic cooling process associated with this system is similar to the absorption process except that the refrigerant is adsorbed onto the surface of the carrier material rather than being taken into chemical solution. The refrigerant is subsequently driven off by heating so it can be condensed and expanded to produce the cooling effect before it is readsorbed onto the carrier to close the cycle.

Rankine Cycle-Inverse Brayton Cycle System: Refrigeration in such a system is obtained by four sequential steps:

(1)   air compression,
(2)   heat removal from the compressed air,
(3)   expansion of the air through a turbine to extract work, and
(4)   discharge of the cool air into the space.

Power to operate this system would have to be obtained from a heat engine such as the Rankine cycle device described previously. Air cycle systems are more commonly used in air conditioning of aircraft than any stationary applications, but their application there has been accompanied by sufficient development to include them within the scope of this discussion.

Night Radiation: Some installations with solar heating utilize nocturnal cooling in the summer. Cooling may be by radiation or evaporation, or both. The best type does not use the collector for radiation because a good collector is designed to be a poor radiator. These systems perform best in very dry climates and high altitudes. Performance is not easy to evaluate because it is, of course, very dependent on local humidity conditions and ambient temperature. It is believed that the usefulness of nocturnal radiation is geographically limited. Where evaporation is used, it will be the forced type presently in use with larger air conditioning installations.

Conclusions: Absorption and Rankine Cycle-Vapor Compression systems are the most promising system concepts. The Absorption system is probably best suited to the lower temperature boiler applications, and RC-VC system best suited to the higher ones.

References

(1) Löf, G.O.G., "The Heating and Cooling of Buildings with Solar Energy," *Introduction to the Utilization of Solar Energy,* McGraw-Hill, New York, 1963, 239-294.

(2) Löf, G.O.G., "Use of Solar Energy for Heating Purposes: Space Heating," *Proceedings of the United Nations Conference on New Sources of Energy — Solar Energy:II,* Rome, August 21-31, 1961, vol 5, New York, 1964, 114-138-139.

(3) Daniels, F., "Introduction to the Utilization of Solar Energy," *Introduction to the Utilization of Solar Energy,* McGraw-Hill, New York, 1963, 1-11.

(4) Hottel, H.C. and Erway, D.D., "Collection of Solar Energy," *Introduction to the Utilization of Solar Energy,* McGraw-Hill, New York, 1963, 87-106.

(5) Whillier, A., "Design Factors Influencing Solar Collector Performance," Chapter III of ASHRAE Bulletin, *Low Temperature Engineering Application of Solar Energy,* 1967.

(6) *Thin Films and Solar Energy,* Brochure by Optical Coating Laboratory, Inc., P.O. Box 1599, Santa Rosa, California.

(7) Kudryashova, M.D., "Mechanical Treatment of Collector Surfaces in Solar Installations Leading to Improved Selectivity of Optical Properties," *Geliotekhnika,* vol 5, No. 5, 1969, 36-39.

(8) Cunnington, G.R. and Streed, E.R., "Experimental Performance of a Honeycomb Covered Flat Plate Solar Collector," Paper presented at Annual Meeting of International Solar Energy Society, U.S. Section, October, 1973.

(9) Edwards, D.K., Gier, J.T., Nelson, K.E. and Reddick, R.D., "Spectral and Directional Thermal Radiation Characteristics of Selective Surfaces for Solar Collectors," ASME Paper S/43, 536-551.

(10) Farber, E., "Selective Surfaces and Solar Absorbers," *Solar Energy,* vol 5, No. 2, 1961, 37.

(11) Tabor, H., "Selective Surfaces for Solar Collectors," Chapter IV of ASHRAE Bulletin, *Low Temperature Engineering Application of Solar Energy,* 1967.

(12) Keller, Arnold, "Selective Surfaces of Copper Foils," Proc. UNESCO International Congress, *The Sun in the Service of Mankind,* Paris, France, Paper No. E-43, July, 1973.

(13) Rosenthal, et al, "Solar Power Prospects," Paper presented at NSF Solar Thermal Conversion Workshop, Jan 11-12, Arlington, Va, 1973, 4-8.

(14) Hollands, "Directional Selectivity, Emittance, and Absorptance Properties of Vee-Corrugated Specular Surfaces," *Solar Energy,* vol 7, No. 3, 1963, 108.

(15) Chiou, J.P., El-Walkil, M.M. and Duffie, J.A., "A Slit and Expanded Aluminum-Foil Matrix Solar Collector," *Solar Energy,* vol 9, Apr.-June, 1965, 73.

(16) Swartman, R.K. and Ogunlade, O., "An Investigation of Packed-Bed Collectors," *Solar Energy,* vol 10, No. 3, 1966, 106.

(17) Bevill, V.D. and Brandt, H., "A Solar Energy Collector for Heating Air," *Solar Energy,* vol 12, No. 1, 1968, 19.

(18) Beckman, W.A., "Radiation and Convection Heat Transfer in a Porous Bed," Paper No. 67-WA/SOL-1, Presented at Winter Annual Meeting of American Society of Mechanical Engineers, November, 1967.

(19) Lalude, D. and Buckberg, H., "Design and Application of Honeycomb Porous-Bed Solar-Air Heaters," *Solar Energy,* vol 13, 1971, 223-242.

(20) Shoemaker, "Notes on a Solar Collector with a Unique Air Permeable Media, *Solar Energy,* vol 5, No. 4, 1961, 138.

(21) Rankine, A.D. and Charters, W.W.S., "Combined Convective and Radiative Heat Losses from Flat-Plate Solar-Air Heaters," *Solar Energy,* vol 12, No. 4, 1968, 517.

(22) Charters, W.W.S. and Peterson, L.F., "Free Convection Suppression Using Honeycomb Cellular Materials," *Solar Energy,* vol 13, July, 1972, 353-361.

(23) Charters, W.W.S., "Heat Flux Measurements for Free Convection Within Inclinal Cellular Structures," Paper E30. Proc. International Congress, *The Sun in the Service of Mankind,* Paris, France, Paris Conference, July, 1973.

(24) Buchberg, H., Edwards, D.K. and Lande, O., "Design Considerations for Cellular Solar Collectors," ASME Paper #68-WA/SOL-3, Presented at Annual Winter Meeting of ASME, 1968.

(25) Buchberg, H., Lalude, D.A., and Edwards, D.K., "Performance Characteristics of Rectangular Honeycomb Solar-Thermal Converters," *Solar Energy,* vol 13, May, 1971, 193-221.

(26) Cobble, M.H., "Heating a Solid by Solar Radiation; Heating a Fluid by Solar Radiation," *Solar Energy,* vol 8, No. 2, 1964.

(27) Tabor, H. "Cellular Insulation (Honeycombs)," *Solar Energy,* vol 12, No. 4, 1968, 549.

(28) Hollands, K.G.T., "Honeycomb Devices in Flat Plate Solar Collectors,"
      Solar Energy, vol 9, No. 3, 1965, 159.

(29) Speyer, E., "Solar Energy Collection with Evacuated Tubes," ASME
      Paper No. 64-WA/SOL-2, 1964.

(30) Satcunanathan, S. and Deonarine, S., "A Two Pass Solar Air Heater,"
      Solar Energy, vol 15, 1973, 41-49.

(31) Löf, G.O.G., "Performance of Solar Energy Collectors of Overlapped-
      Glass-Plate Type," Library of Dr. G.O.G. Löf, Colorado State Univer-
      sity, Ft. Collins, Colorado.

(32) Selcuk, K. "Thermal and Economic Analysis of the Overlapped-Glass
      Plate Solar-Air Heater," *Solar Energy,* vol 13, May, 1971, 165-191.

(33) Holman, J.P., *Heat Transfer,* Third Edition, McGraw-Hill Book Co., 1972.

(34) Charters, W.W.S. and Macdonald, R.W.G., "Heat Transfer Effects in Solar
      Air Heaters," Library of Dr. G.O.G. Löf, Colorado State University,
      Fort Collins, Colorado.

(35) Close, D.J., "Solar Air Heaters," *Solar Energy,* vol 7, No. 3, 1963, 117.

(36) Gupta, C.L. and Garig, H.P., "Performance Studies on Solar Air Heaters,
      Solar Energy, vol 11, No. 1, 1967, 25-31.

(37) Belton, G. and Ajami, F., *Thermochemistry of Salt Hydrates,* Report No.
      NSF/RANN/SE/GI 27976/TR/73/4, University of Pennsylvania,
      Philadelphia, 1973.

(38) Kauffman, K. and Pan, Y.C., *Thermal Energy Storage in Sodium Sulfate
      Decahydrate Mixtures,* Report No. NSF/RANN/SE/GI 27976/TR/72/11,
      University of Pennsylvania, Philadelphia, 1972.

(39) Altman, M, *Conservation and Better Utilization of Electric Power by
      Means of Thermal Energy Storage and Solar Heating, Phase II — Progress
      Report,* Report No. NSF/RANN/SE/GI 27976/TR/72/4, University of
      Pennsylvania, Philadelphia, 1972.

(40) Whillier, A., *Solar Energy Collection and Its Utilization for House Heating,*
      Doctor of Science thesis, Dept. of Mechanical Engineering, Massachusetts
      Institute of Technology, Cambridge, Massachusetts, 1953.

(41) Speyer, E., "Optimum Storage of Heat with a Solar House," *Solar Energy,*
      vol III, No. 4, December, 1959, 24-48.

Chronology of Experimental Systems

CHRONOLOGY

During the last 35 years, about 30 solar-powered systems have been developed throughout the world for space heating and cooling, and considerably more have been developed for domestic water heating. The solar-powered heating and cooling systems have ranged from simple passive systems where solar heat is absorbed in a massive concrete wall and distributed to the area of use by convection, to sophisticated automated systems employing fused salt storage, heat pumps, absorption air conditioners, and solar cell electrical generators.

A review of these systems is given in PB-235 428, *Solar Heating and Cooling of Buildings, Phase O, Final Report, Volume II (Appendices A-N)* prepared by Westinghouse Electric Corporation for the National Science Foundation, May 1974.

Table 2.1 lists some of the most important subsystem parameters and system performance characteristics for a number of these existing systems where sufficient information was available. It also includes preliminary information on a number of systems presently under development. Much of the information contained in Table 2.1 has been compiled previously by the University of Pennsylvania (1) and others in the field (2), (3), (4), (5), (6), (7), (8), (9), (10).

All of the listed systems are characterized by individual custom design, custom fabrication, and high capital cost. Several of these systems have been well instrumented, and considerable data characterizing their performance has been accumulated. Direct comparison of performance among the various systems and subsystems in use is quite difficult, however, because of the wide variation in system configurations, differences in system climatic environments, and differences in the performance parameters measured.

(continued)

TABLE 2.1: SOLAR HEATING AND COOLING SYSTEMS

Item No. Building Type Size	Location Latitude Degree Days	Collector Size Tilt, Efficiency Construction	Storage Type Size, Material Capacity	Transport Fluid Collector to Storage	Storage to Load	Energy Use Auxiliary Source Remarks
1. M.I.T. House #1 Laboratory 46.5 m²	Cambridge, Mass. 42° N 5800 D.D.	38 m² 30°, - 3 Glass Covers	Sensible Heat 66,000 kg Water	Water	Air	Space Heating - Heat Loss = 3.1 kw @ 18° C
2. M.I.T. House #2 Laboratory 65 m²	Same	21 m² Vertical Wall 7 Different Types, 3 m²	Various Including No Storage	Various	Air	Space Heating - Converted to #3
3. M.I.T. House #3 Dwelling 56 m²	Same	37 m² 57°, - Cu Tube & Plate 2 Glass Covers	Sensible Heat 4,500 kg Water	Water	Air	75% Solar Heating Oil Auxiliary Heat Loss=5.9 kw @ -18° C
4. M.I.T. House #4 2 Story Dwelling 135 m²	Same	59 m² 60°, 43% Al Plate & Cu Tube 2 Glass Covers	Sensible Heat 5,700 kg Water plus 1,000 kg Water Heated by Oil	Water	Air	48% Solar Heating Oil Auxiliary Heat Loss=8.8 kw @ -18° C
5. Löf House 93 m²	Boulder, Colorado 40° N 5500 D.D.	43 m² 27°, - Black Glass Plates	Sensible Heat 7,300 kg Rock	Air Can be combined		20-25% Solar Heat Nat. Gas Aux. Insufficient Insulation
6. USFS Sta. Amado Dwelling 62 m²	Amado, Arizona 32° N 525 D.D. (Jan.)	29 m² 53°, - Black Cloth Screen 1 Glass Cover	Sensible Heat 59,000 kg Rock 1.4×10^9 J @ 32 - 60° C	Air @ 60° C Can be combined		100% Solar Heat No aux. needed System overdesigned, cooling by night, Air & Rad to -17° night sky

TABLE 2.1: (continued)

Item No. Building Type Size	Location Latitude Degree Days	Collector Size Tilt, Efficiency Construction	Storage Type Size, Material Capacity	Transport Fluid Collector to Storage	Transport Fluid Storage to Load	Energy Use Auxiliary Source Remarks
7. Office Building (Single Story) 400 m²	Albuquerque, N.M. 35° N. 4400 D.D.	73 m² 60°, - Al Plate & Cu Tube 1 Glass Cover	Sensible Heat 23,000 kg Water	Water	Water @ 32-43° C in Floor & Ceiling	Space Heating Heat Pump Aux. Heat Pump Evapo. Cooling to Air
8. Denver House Löf Dwelling 297 m²	Denver, Colorado 40° N 5700 D.D.	49 m² 45°, 35% Black Glass Plates 2 Glass Covers	Sensible Heat 10,900 kg Rock 0.32×10^9 J	Air @ 80° C May be Combined	Air	26% Solar Heat Nat. Gas Aux. Domestic Water Preheating
9. AFASE House Dwelling -	Phoenix, Arizona 33° N 1500 D.D.	65m² Variable Tilt - Motor Driven Cu Plate & Cu Tube, Plastic Cover	Sensible Heat 7,600 kg Water	Water	Air (via heat pump)	Space Heating Heat Pump is part of solar sys (not aux), provides cooling- uses Swimming Pool as Heat Sink
10. University of Arizona Solar Laboratory 149 m²	Tucson, Arizona 32° N 1800 D.D.	151 m² 15°, 29% Green Cu Sheet & Cu Tubes, No Glass	Sensible Heat 17,000 kg Water in 2 Sections	Water	Water @ 43° C in Ceilings	86% Solar Heat Heat Pump part of Solar System Transfers Heat between Storage Tanks - Summer Night Sky Radiation of $5.3(10^8$ J for Cooling
11. Thomason House #1 Dwelling 139 m²	Washington, D.C. 39° N 4300 D.D.	78 m² 60° 45- Corrugated Iron Sheets, 1 Glass, 1 Plastic Cover	Sensible Heat 6,000 kg Water 45,000 kg Rock 1.7×10^9 J	Water @ 57° C	Air	95% Solar Heat(Heat) Oil Aux. Cooling by Evaporation & Night Sky Radiation

(continued)

TABLE 2.1: (continued)

Item No. Building Type Size	Location Latitude Degree Days	Collector Size Tilt, Efficiency Construction	Storage Type Size, Material Capacity	Transport Fluid Collector to Storage	Transport Fluid Storage to Load	Energy Use Auxiliary Source Remarks
12. Dover House Dwelling 135 m²	Dover, Mass. 42° N 6,000 + D.D.	67 m² 90°, 41% Black Iron Sheet 2 Glass Covers	Latent Heat 19,000 kg Glauber Salt 5×10^9 J Adequate for 5-7 Cloudy Days	Air	Air	Can be Combined; 100% Solar Heat First Season - Less Later Seasons
13. Solar House 1-1/2 Story 102 m²	State College, N.M. 32° N	42 m² 45°, - Black Steel Sheet 1 Glass Cover	Latent Heat 1,800 kg Glauber Salt 0.53×10^9 J	Air	Air	50% Solar Heat - Collected 25% more than needed in December.
14. Solar Laboratory Single Story 111 m²	Princeton, N.J. 40° N 5100 D.D.	56 m² 90°, 46% Black Metal Sheet 2 Glass Covers	Latent Heat 11,000 kg Glauber Salt 2.6×10^9 J	Air	Air	100% Solar Heat No Aux.
15. Solar Heated House	New Haven, W.Va. 39° N	10.4 m² 50°, - Heat Pump Evaporator Plate 1 Glass Cover	None	Heat Pump		Collected 60% of Insolation - on Warm days heat collected may exceed Insolation.
16. - Solar Heated House 229 m²	Tokyo, Japan 36° N	131 m² 15°, 22% Corrugated Sheet Metal - No Cover	Sensible Heat 36,000 kg Water 10,000 kg Water in two tanks	Water @ 27° C Max.	Water in Ceiling	42% Solar Heat Heat Pump Aux. Cooling of Storage by Evaporation & Night Sky Radiation

(continued)

TABLE 2.1: (continued)

Item No. Building Type Size	Location Latitude Degree Days	Collector Size Tilt, Efficiency Construction	Storage Type Size, Material Capacity	Transport Fluid Collector to Storage	Transport Fluid Storage to Load	Energy Use Auxilliary Source Remarks
17. Solar Heated Laboratory 82 m²	Nagoyo, Japan 35° N	28 m² 35°, 22% Corrugated Sheet Metal - No Cover	Sensible Heat 11,000 kg Water in Two Tanks	Water @ 27° C Max.	Air	47% Solar Heat Heat Pump Aux. Cooling of Storage by Evaporation & Night Sky Radiation
18. Laboratory 180 m²	Capri, Italy 41° N	30 m² 90°, - Metal Plate in Walls - 1 Glass, 1 Plastic Cover	Sensible Heat 3,000 kg Water	Water	Water	50 - 70% Solar Heat - Electric Aux.
19. Solar Heated House 75 m²	Odeillo, France ≈ 43° N	48 m² 90°, - Concrete Wall 2 Glass Covers	Sensible Heat 91,000 kg Concrete Collector Wall	Integral	Air	50% Solar Heat Passive System uses Convection for circulation. May also use for circulation of outside air for cooling
20. CSIRO Office and Laboratory	Melbourne, Australia 38° S	56 m² - Corrugated Iron Sheet	Sensible Heat 32 m³ Rock	Air	Air	Space Heating & Cooling Rock Storage is cooled at night by evaporative cooler

(continued)

TABLE 2.1: (continued)

Item No. Building Type Size	Location Latitude Degree Days	Collector Size Tilt, Efficiency Construction	Storage Type Size, Material Capacity	Transport Fluid Collector to Storage	Storage to Load	Energy Use Auxiliary Source Remarks
21. Brisbane Solar House 123 m²	Brisbane, Australia 27° S	6 m² 10°, 20-40% Cu Tube & Plate with selective Black Surface 2 Glass Covers	Sensible Heat 265 kg Water 27,000 kg Rock	Water	Air	100% Cooling* by Li-Br-H₂O Air Conditioner projected for 65 m² Collector for Basis of Measurements
22. Delaware Solar House 149 m²	Newark, Del. 40° N	75 m² @ 45° Tilt, 9 m² @ 90° Al Sheet & CdS & Cu₂S Solar Cells 2 Glass Covers	Latent Heat Eutectic Salts Melting Points 10°C, 20°C, 40°C 110 V. DC Lead-Acid Storage Battery	Air	Air	Heating, Cooling & Electricity Heat Pump Aux. Cooling of Main Storage (20°C Salt) by Night Sky Radiation
23. University of Florida Solar House	Gainesville, Fla. 30° N	33 m² 45°, - Cu Sheet & Tube 1 Glass Cover	Sensible Heat 11,400 kg Water	Water	Water	100% Heating, no aux. Separate water htr. & solar heated swim.pl. Solar cells for elect.
24. Zomeworks House	Albuquerque, N.M. 35° N 4400 D.D.	- Vertical Wall of Water-Filled Oil Drums - Insolation Shut- ter controlled	Sensible Heat Water-Integral with Collector	None	Air	Provides Heat Source/Sink for Heating & Cooling Passive System
25. Sky Therm. Pond Roof House	Phoenix, Arizona 33° N	Water Pond on Roof Movable, Insu- lated Covers	Sensible Heat Water - Integral with Collector	None	Air Convec- tion & Radia- tion	Heating & Cooling - Passive System

(continued)

TABLE 2.1: (continued)

Item No. Building Type Size	Location Latitude Degree Days	Collector Size Tilt, Efficiency Construction	Storage Type Size, Material Capacity	Transport Fluid		Energy Use Auxilliary Source Remarks
				Collector to Storage	Storage to Load	
26. Sky Therm Pond Roof House	Atascadero, Calif. 35° N	water Pond on Roof Movable, Insulated Covers	Sensible Heat Water - Integral with Collector	None	Air Convection & Radiation	Heating & Cooling - Passive System
27. Thomason "Sunny South" House	Proposed new system - not implemented yet	Water Pond on Roof Clear Plastic Cover, Vertical Reflector	Sensible Heat "Pancake" Water Tank under floor	Water	Air convection & Radiation	Heating & Cooling - Can remove cover in Summer for evaporative cooling as well as Night Sky Radiation
28. Massachusetts Audubon Society Headquarters (Proposed) 744 m²	Lincoln, Mass. 42° N	325 m² 45°, - Black Absorber 2 Glass Covers	Sensible Heat 28,000 kg Water	Water	Air	Estimated 65-75% Solar Heating
29. A.N. Wilson House (Proposed) 130 m²	Shanghai, W.Va. 39° N	55 m² 45°, - Tube and Plate 1 Glass, 2 Mylar Covers	Sensible Heat 6,400 kg Water @ 32-54° C. & 4,000 kg Water @ 57-66° C	Water and 50% ethylene glycol	Air	Est. 18% of heating & Electrical Load provided by Solar System (inc. Solar Cell System)

Source: PB-235 428, May 1974

SYSTEM CONFIGURATIONS

Basically, solar-powered heating and cooling systems can be characterized according to the working fluid used to transport the heat generated by the collection of solar energy. The two most widely used working fluids thus far are air and water. Both have the advantages of universal availability, low cost, non-toxicity, and widespread use in conventional heating and cooling systems.

These factors, together with the inherent capability of interfacing and integrating air and water subsystems with either new or existing conventional heating and cooling systems, make them the most practical working fluids for now and the near future. New materials, such as organic fluids, or other technical advances may provide better working fluids at some future time, but this study will consider only air and water as the basic working fluids. In some instance, additives, such as antifreeze or corrosion inhibitors, may be used to improve the basic properties of water as a working fluid.

Systems Using Air as the Working Fluid

Systems using air as the working fluid usually employ flat plate collectors to convert the solar energy to heat, which is then carried by the air, at temperatures ranging to 140°C to a point of heat storage, use, or heat exchange.

Heat storage for air systems may be a container of rocks for sensible heat storage or a stack of containers of fusible salts for combined sensible and latent heat storage. The majority of solar-powered systems employing air as the working fluid provide only space heating from the solar energy collected. (See Table 2.1.) Of the air transport systems listed, three provide cooling, although not by heat-operated refrigeration systems, and only one provides domestic water heating.

Space Heating: Solar-powered systems using hot air transport interface quite simply with conventional forced-air space heating systems in new construction. A flat plate collector, a heat storage unit, some dampers, and controls are all that are needed in addition to the conventional forced-air subsystems.

Retrofit of hot air solar systems to existing buildings, however, may not prove to be so simple. Mounting of the solar collector on an existing roof or wall may require structural modification and still not provide optimum collector orientation. Space required for the heat storage unit, which is usually large and heavy, may also entail some modification to the building structure. Finally, it may be difficult to add the required ducting to the existing structure without extensive modification.

A typical solar-powered space heating system employing air heat transport is shown in the block diagram of Figure 2.1. Air is heated in the flat plate solar collector, then either circulated through the space to be heated, or drawn through the heat storage medium. Preheating of the domestic hot water supply is also provided.

FIGURE 2.1: SOLAR-POWERED HOT AIR SPACE HEATING SYSTEM

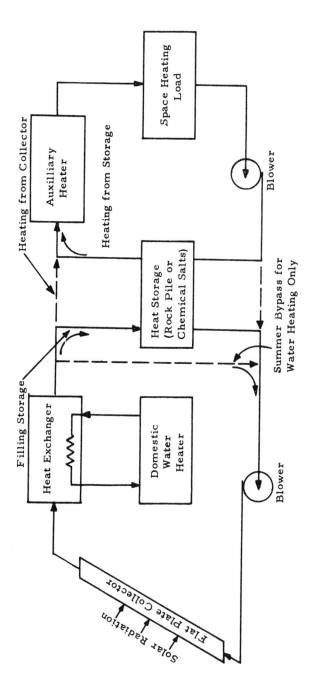

Domestic Water Heating: The only example of domestic hot water heating from a hot air solar system is the Löf house [#5 in Table 2.1, (3)]. A finned tube heat exchanger between the solar collector and the heat storage provides pre-heating of the water supplied to an automatic gas-fired water heater. In the summer when it is not desirable to store heat, the storage is bypassed and the collected solar heat is used for water heating only. An 80 gallon tank in the domestic hot water system provides some storage of the solar heat.

Space Cooling: Hot air solar systems have not been widely used in applications requiring space cooling. The simpler systems which cool the storage medium by air circulated through an evaporative cooler or through the collector for night sky radiative cooling have not generally proven so effective as conventional air conditioners.

One exception seems to be the system employed in the CSIRO office and laboratory in Melbourne, Australia (#20 in Table 2.1). In the CSIRO system, the rock pile storage is cooled by an evaporative cooler at night when the wet bulb temperature is low, and dry air is circulated through the rock pile next day to cool the building. The CSIRO system has provided very satisfactory performance for 10 hours per day during the summer (8).

Interfacing absorption type air-conditioners or heat pumps with hot air solar systems is generally more difficult than with hot water solar systems. Of the hot air systems listed in Table 2.1, none employ absorption air-conditioners and only one employs a heat pump. The Delaware House (#22 in Table 2.1) uses an electrically driven heat pump which operates at night to freeze a eutectic salt with a 10°C melting point (6), (7). During the day, room air is circulated through the cold salt reservoir to provide space cooling. On clear nights, air may be circulated through the collector to use radiation cooling of the main reservoir of eutectic salt with a melting point of 20°C. This reservoir then provides a cooler sink for the heat pump condenser to enhance the efficiency of its refrigeration cycle.

Systems Using Water as the Working Fluid

Systems using water as the working fluid usually employ flat plate collectors where the water is circulated through a series of tubes in intimate contact with the collector plate. The heated water (at temperatures ranging to approximately 100°C) is then returned to an insulated tank for storage. Where there is danger of freezing, the water in the collector loop may contain an antifreeze solution such as ethylene glycol. To reduce the quantity of antifreeze required (and hence the cost), a heat exchanger is frequently used between the collector loop and the storage water.

Although it is possible to use latent heat storage in hot water systems through the addition of chemical salts (either directly or in sealed containers), all of the hot water systems listed in Table 2.1 use sensible heat storage.

The complexity of interfacing solar powered subsystems using hot water transport with conventional heating and cooling subsystems will depend largely on
the performance objectives and mode of usage of the overall system design. As
in the case of the hot air transport solar systems, retrofit to existing buildings
may require structural modifications to accommodate the solar collectors and
heat storage. Addition of water piping to existing structures is usually somewhat easier than adding air ducts because of the smaller space required. Figure 2.2 is a block diagram depicting one method of providing space heating,
space cooling and domestic water heating with solar energy.

FIGURE 2.2: SOLAR HOT WATER HEATING AND COOLING SYSTEM

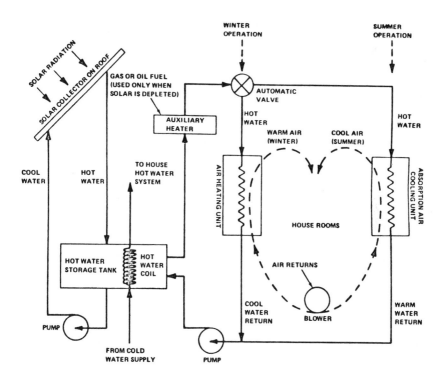

Space Heating: In the system shown in Figure 2.2, solar radiation is converted
to heat in a flat plate collector and carried by circulating water to a tank where
it is stored as sensible heat. When space heating is required, hot water from
storage is pumped through a heat exchanger where it heats circulating room air.
If sufficient heat is not available from storage, additional heat is provided by
the auxiliary heater burning gas or fuel oil.

An alternative system which employs a heat pump to amplify the heat provided by the solar system is shown in Figure 2.3. This system configuration is employed in the University of Arizona solar laboratory (#10 in Table 2.1), the solar heated laboratory at Nagoyo, Japan (#17 in Table 2.1), and in the Yanagi-machi solar heated house in Tokyo, Japan (#16 in Table 2.1).

The heat pump extracts heat from the water in storage tank #1 even when it is several degrees below temperatures useful for direct space heating. Storage tank #2 permits operation of the heat pump during off peak hours. Hot water from storage tank #2 is circulated through radiant ceiling panels to heat the rooms through infrared radiation and natural convection.

Domestic Water Heating: In the system shown in Figure 2.2, cold water from the house supply is passed through a coil immersed in the solar heat storage tank and into a conventional water heater where additional heat is added as required.

Figure 2.4 shows two configurations for domestic water heating with solar energy. Both are completely independent systems and thus they may be used alone where only domestic water heating is required, or may be used in conjunction with a separate solar space heating and/or cooling system.

The configuration in Figure 2.4a uses a pump to circulate the hot water from the collector through the heat exchange coil in the storage tank. .With this configuration, an antifreeze solution may be used in the collector loop to prevent freezing. With a properly designed heat exchanger, performance will not be significantly less than that of the systems which circulate the domestic water directly through the collector (12). By placing the bottom of the storage tank approximately a foot higher than the collector, natural circulation of the water through the collector is provided by thermosyphon action (12), as shown in Figure 2.4b. However, except for relatively small tanks, such mounting is usually not desirable.

Space Cooling: In the system shown in Figure 2.2, summer cooling is provided by an absorption air-conditioning unit. Hot water from storage provides heat for the generator of the absorption unit at temperatures ranging from approximately $50°$ to $100°C$ (5), (9), (11). Water below about $80°C$ requires supplementary heat before it can efficiently drive the air-conditioner. An alternative method of using an absorption air-conditioner is shown in Figure 2.5.

Hot water from the collector is fed through a mixing valve where it is tempered with cold water if necessary to prevent excessive temperatures which might crystallize the absorbent out of solution, and then to the generator of the absorption air-conditioner. Water in a storage tank is chilled by the evaporator of the air-conditioner. The chilled water is then circulated through a heat exchanger (or radiant panels) as needed to provide cooling of the rooms. This system has the advantage that the absorption air-conditioner is operating from the higher temperatures supplied from the collector (as compared to storage temperatures) and hence can have a higher coefficient of performance.

FIGURE 2.3: SOLAR HEATING AND COOLING SYSTEM WITH HEAT PUMP

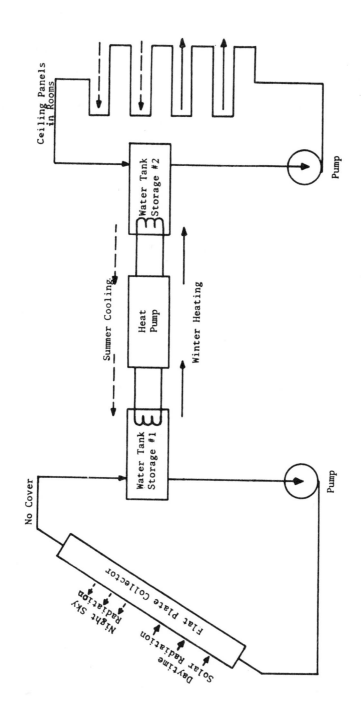

FIGURE 2.4: DOMESTIC WATER HEATING WITH SOLAR ENERGY

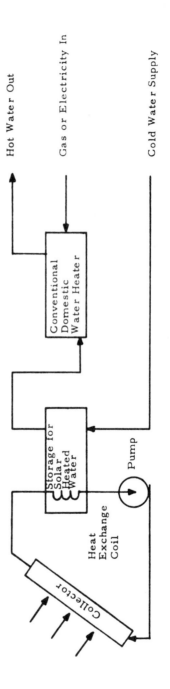

Water Heating System with Circulating Pump

(continued)

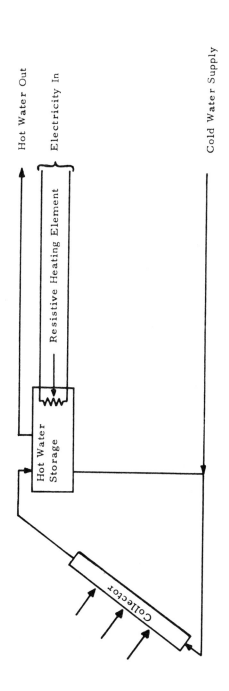

Hot Water Out

Electricity In

Resistive Heating Element

Cold Water Supply

Hot Water Storage

Collector

Water Heater with Thermosyphon Circulation

FIGURE 2.4: (continued)

FIGURE 2.5: ALTERNATIVE USE OF ABSORPTION AIR CONDITIONER

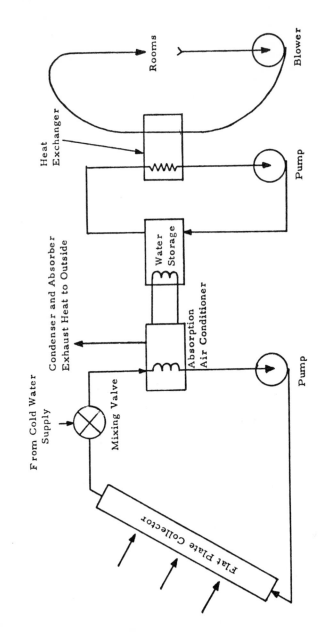

The systems typified by Figure 2.3 use a heat pump during off peak hours to chill the water in storage tank #2. The cooled water is then circulated through the radiant ceiling panels as needed to cool the rooms. The heat transferred into storage tank #1 by the heat pump may be discharged by the coverless collector through convection and night sky radiation. A more widely applicable alternative is cooling by forced evaporation in a cooling tower.

Systems with Integral Collector and Storage

Another class of systems which utilize solar energy for space heating of buildings provides natural air-conditioning through the use of integral collector/storage units which act as large thermal time constants to average diurnal temperature variations. Included in this category are the Sky Therm (pond roof) houses in Phoenix (#25 in Table 2.1) and Atascadero (#26 in Table 2.1), the Zomeworks house (water-filled drum wall) in Albuquerque (#24 in Table 2.1) and the solar heated house (large concrete wall) in Odeillo, France (#19 in Table 2.1).

All of these systems are basically passive as they collect and store the solar heat in the same element for later space heating through radiation and natural convection.

In the Sky Therm system, roof top water ponds with black plastic bottom linings are exposed to the winter sun to accumulate solar heat during the day (3). At night movable insulation is positioned over the ponds to retain the accumulated heat. In the summer, the insulation is positioned over the ponds during the day to exclude the incident solar radiation. During the night, the insulation is removed and the pond water is cooled by evaporation and radiation to the night sky.

The water-filled drum wall of the Zomeworks system is exposed to insolation from the winter sun to accumulate solar heat during the day (14). At night, shutters are closed to prevent loss of heat to the outdoors and the accumulated energy heats the room through radiation and natural convection. During the summer, the drum wall can be covered during the day to exclude insolation, then exposed to night air for convective and radiative cooling.

The solar heat house at Odeillo, France, employs a massive blackened concrete wall behind a large glass window to collect the solar heat. Cool air from the room is allowed to enter the space between window and the wall where it is heated. The heated air rises and is ducted back into the room, thus providing circulation by natural thermal convection (15).

A three-unit dwelling under construction at Odeillo will utilize a similar system for winter solar heating. In addition, it will have provision for moderate summer cooling by allowing cool air from the shaded north side of the building to enter the room. Room air is carried by natural convection up through the space between the glass and the concrete wall and is exhausted to the outside through a damper at the top.

Other Configurations

A system described by Thomason (10), which he calls the Solaris Sunny South Model, uses a rooftop solar collector pond similar to that of the Sky Therm System. Water in the rooftop pond is heated by the sun in the daytime and then allowed to drain into a pancake storage tank under the building floor where it provides nighttime heating of the rooms by radiation and natural convection. A small pump is used to transfer the water from storage into the rooftop pond.

Moderate summer cooling can be provided by evaporation and night sky radiation from the pond. Some water may be left in the pond during the day to reduce ceiling heating.

A passive solid adsorbent air-conditioner has been implemented by Dannies (16) in Liberia. It consists of banks of solid adsorbent mounted in channels in the east and west walls of a building. In the morning, the sun heats the adsorbent collectors on the east wall, creating natural air convection from inside the building up through the adsorbent to the outside. The exhaust air removes moisture from the heated adsorbent, thus regenerating it.

Simultaneously, outside air is drawn in from the opposite side of the building through the other shaded set of adsorbent collectors, where it gives up its moisture. As it enters the room, the dry air passes over a water bath, where it is cooled by evaporation. In the afternoon, the west side of the building is heated by the sun and the process flows in the reverse direction.

Russian solar scientists have combined a similar solid adsorbent air drying system with an air-to-air heat exchanger to improve the efficiency of an evaporative air cooling system (17).

References

(1) Lorsch, H.G., *Solar Heating Systems Analysis,* submitted to NSF under Grant No. GI-27967 by University of Pennsylvania, November, 1972.

(2) Löf, G.O.G., "Use of Solar Energy for Heating Purposes: Space Heating," *Proc. U.N. Conf. on New Sources of Energy,* Rome, 5, 114, Paper No. GR/14(s), 1961.

(3) Löf, G.O.G., El-Wakil, M.M. and Chiou, J.P., "Design and Performance of Domestic Heating System Employing Solar Heated Air—The Colorado Solar House," *Proc. U.N. Conf. on New Sources of Energy,* Rome, 5, 185, Paper No. 5/114, 1961.

(4) Löf, G.O.G. and Ward, D.S., "Solar Heating and Cooling: Untapped Energy Put to Use," *Civil Engineering—ASCE,* Sept., 1973, pp 88-92.

(5) Farber, E.A. et al, "The University of Florida Solar House," *Proc. UNESCO Internat. Congress, The Sun in the Service of Mankind,* Paris, Paper No. EH 107, July, 1973.

(6) Böer, K.W., "Combined Solar Thermal and Electrical House System," *Proc. UNESCO Internat. Congress, The Sun in the Service of Mankind,* Paris, Paper No. EH 108-1, July, 1973.

(7) Böer, K.W., "The Solar House and Its Portent," *Chemtech,* pp 394-400, July, 1973.

(8) Close, D.J., Dunkle, R.V. and Robeson, K.A., "Design and Performance of Thermal Storage Air Conditioning Systems," *Instn. Engrs., Australia—Mech. & Chem. Eng. Trans.,* Vol. MC4, No. 1, pp. 45-54, May, 1968.

(9) Sheridan, N.R., "Performance of the Brisbane Solar House," *Solar Energy,* Vol. 13, 1972, pp. 395-401.

(10) Thomason, H.E. and Thomason, H.J.L., Jr., "Solar Houses/Heating and Cooling Progress Report," *Solar Energy,* Vol. 15, 1973, pp. 27-39.

(11) Duffie, J.A., Chung, R. and Löf, G.O.G., "A Study of a Solar Air Conditioner," *Mechanical Engineering,* pp. 31-35, August, 1973.

(12) Löf, G.O.G. and Close, D.J., "Solar Water Heaters," ASHRAE Handbook, *Low Temperature Engineering Application of Solar Energy,* Chapter VI, pp. 61-78, 1967.

(13) Hay, H.R. and Yellott, J.I., "International Aspects of Air Conditioning with Movable Insulation," *Solar Energy,* Vol. 12, 1969, pp. 427-438.

(14) Baer, S., "The Drum Wall," *Proc. Heating and Cooling for Buildings Workshop,* Washington, D.C., Session VII, pp. 186-187, 1973.

(15) Walton, J.D., Jr., "Space Heating at the C.N.R.S. Laboratory, Odeillo, France," *Proc. Solar Heating and Cooling for Buildings Workshop,* Washington, D.C., Session VI, pp. 127-139, 1973.

(16) Dannies, I.H., "Solar Air Conditioning and Solar Refrigeration," *Solar Energy,* Vol. 3, No. 1, 1959, pp. 34-39.

(17) Akopdzhanyan, E.S., "Solar Air Conditioning with Solid Absorbents, Applied Solar Energy," *Geliotekhnika,* Vol. 3, No. 6, 1967, pp. 81-85.

Descriptions of Experimental Systems

SOLAR HEATING SYSTEMS

The status of solar heating developments will be outlined in very general form, with respect to the design and operation of several completed solar-heated structures. These applications should be considered as in early experimental stages, since to date only a few dozen solar-heated structures have been built and tested. A solar-heated structure may be defined as a building in which a solar energy collector in combination with a heat-storage unit provides a substantial portion of the heat required in the structure. Since there are substantial differences in the design and operating characteristics of practically all the existing solar heating systems, brief descriptions of each are presented here. Where the data are available, performance is given in brief. Several structures originally provided with solar heating systems, although they are no longer in service, are also described because the designs are distinct from those now operating.

The information in this chapter was adapted from AD-778 846, *Solar Energy* by V.A. Stevovich of Informatics, Inc., prepared for the Air Force Office of Scientific Research Advanced Research Projects Agency and issued in March 1974; PB-223 536, *Proceedings of the Solar Heating and Cooling for Buildings Workshop, Held in Washington, D.C. on March 21–23, 1973,* edited by R. Allen of the University of Maryland, prepared for the National Science Foundation and issued July 1973 (also available as NSF-RA-N-73-004); and PB-235 432, *Solar Heating and Cooling of Buildings, Phase 0 Feasibility and Planning Study, Final Report, Volume II, Technical Report* by General Electric Company prepared for the National Science Foundation, issued May 1974.

MIT Lexington, Mass. House

In 1958, a solar-heated residence was completed at Lexington, Mass. by MIT (1).

The heating system comprises a solar water heater mounted on a large sloping roof, hot-water storage, oil-fired auxiliary furnace, and hot-air supply to the house rooms. General features of the heating system are shown in Figure 3.1. The house has two floors of living space totaling 1,450 ft². The south elevation is exclusively occupied by a 640 ft² solar collector in 60 panels, each 4 ft long and 32" wide. The roof is oriented at an angle of 60° to the horizontal for ideal exposure to the winter sun. Solar energy is absorbed on a blackened aluminum sheet to which are fastened ⅜" copper tubes 5" apart.

FIGURE 3.1: FUNCTIONAL DIAGRAM OF LEXINGTON, MASS. HOUSE

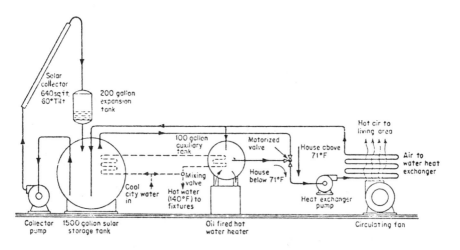

Source: AD-778 846, March 1974

The absorbing surface is backed by glass fiber insulation and covered by 2 layers of window glass spaced ¾" apart. Heat storage is provided in a 1,500-gallon tank from which water is pumped to a heat exchanger for heating air circulated from the house by a fan. An oil-fired water heater provides a reserve of hot water (135° to 150°F) in a 275 gallon-tank. This tank is automatically called upon to deliver hot water to the heat exchanger whenever the temperature in the large storage tank is too low to provide adequate heat to the house. After passage through the heat exchanger, water returns to the large storage tank for reheating in the solar collector. Domestic hot water is provided by circulating it through coils in the large and small storage tanks.

Löf Boulder, Colorado House

A solar heating system was adapted to an existing bungalow in Boulder, Colorado, in 1947. The general features were a 463-ft² solar air heater sloped at 27° on the roof of a 1,000-ft² bungalow. Solar heat was stored in 8 tons of

¾" gravel in a horizontal bin. Auxiliary heat was provided by a gas furnace. Results of operation during one winter season indicated an approximate fuel saving of 20 to 25% (1).

Amado, Arizona House

Another solar experiment was the United States Forest Service solar-heated house at Amado, Arizona, about 30 miles south of Tucson. This system was completed in January, 1955, and proved equal to the design requirement that no auxiliary heat be used. It was possible to provide sufficient solar collector surface and heat-storage capacity to supply all the heat demanded by this small 672-ft^2 house in the comparatively mild climate of southern Arizona. The solar collector was built on the ground immediately south of the house and tilted at an angle of 53° with the horizontal. It is 34 ft long, 10 ft high, and has an exposed net area of 315 ft^2. It was of the screen type, several layers of loosely woven black cloth being stretched across the panels beneath one cover glass. Solar energy is absorbed by these black screens, and air circulated through the screens is heated by contact with them. The heat-storage unit comprised an underground concrete bin containing 65 tons of 4" field rocks (1).

Albuquerque, N.M. One-Story Office Building

A one-story office building having a floor area of 4,300 ft^2 was completed in Albuquerque, New Mexico, in 1956 and houses Bridgers and Paxton Consulting Engineers, Inc., the designers. Figure 3.2 is a schematic of its solar heating system.

FIGURE 3.2: SOLAR HEATING WITH AN AUXILIARY HEAT PUMP
ALBUQUERQUE, N.M.

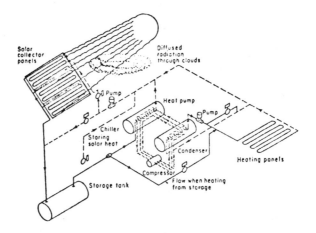

Source: AD-778 846, March 1974

A sloping south wall contains 700 ft^2 of solar collecting panels of the water-heating type. Solar radiation is absorbed on blackened aluminum collector plates containing integral tubing spaces. The collector was subsequently modified by installing copper tubing in contact with the aluminum plate, thereby avoiding corrosion in the aluminum system. The collector is tilted 60° from the horizontal and is glazed with one layer of common window glass.

Heat storage is provided in a 6,000-gallon insulated water tank installed underground. Heat is supplied to the building by circulating warm water from the storage tank to tubing in floor and ceiling panels at relatively modest temperatures of 95° to 110°F. Auxiliary heat is furnished by a heat pump. Summer cooling is provided by a 7½-ton heat pump operating in conventional manner with heat discarded to the atmosphere in an evaporative cooling tower. An interesting supplementary feature is the use of the evaporative water cooler to provide chilled water for circulation through the panel coils whenever the cooling load is moderate (1).

Löf Denver, Colorado House

A 9-room residence in Denver and its associated solar heating system were completed early in 1958 (1). The house is of contemporary architectural style, flat-roofed, one-story (with partial basement), having 2,050 ft^2 of living space on the main floor. The house was designed independently of its solar heating system, but the latter was incorporated into the plans in their final stages. This experimental project was directed by George O.G. Löf for the American-Saint Gobain Corporation (Figure 3.3)

FIGURE 3.3: SCHEMATIC OF DENVER SOLAR HEATING SYSTEM

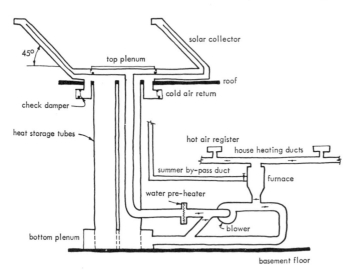

Source: Löf, El-Wakil and Chiou (14)

The solar collector is an air-heating type with the overlapped glass-plate construction. Two 300 ft^2 panels are mounted on the roof at a 45° angle. Each panel contains 20 sections, 6 ft long and 30" wide. Recirculated air passes through two sections in series, once upward and once downward, into a manifold which connects all section exits. Hot air at temperatures up to a maximum of about 175° passes down to the basement, where it first gives up a small amount of heat to the house hot-water supply in a finned heat exchanger connected to an 80-gallon water-storage tank by gravity circulation. The hot air then passes to the suction side of the blower thence either to the bottom of the storage unit or up through the furnace to the distribution system.

The heat storage unit consists of 2 cylinders of fiberboard (standard forms for round concrete columns) filled with approximately 11 tons of 2 to 2½" gravel. The rock column is 18 ft high, extending from the basement floor to the roof. Heat is transferred from air to rocks, the cool air leaving the top of the gravel columns and returning to the solar collectors for reheating. Essentially complete heat transfer is obtained by the large surface area in the storage chambers.

When thermostats in the house call for heat, air is circulated to the rooms directly from the solar collector if the sun is adequate for actuating that particular circulation arrangement. Whenever the solar collector is not receiving solar radiation, the house thermostats cause 2 motorized dampers to shift their position, permitting the blower to draw air down through the heated storage chambers and force it through the furnace to the house distribution ducts. Cold-air ducts return air from the rooms to the top of the storage unit for reheating. Auxiliary heat is supplied to air in transit to the rooms by combustion of natural gas in a furnace through which the air passes. A thermostatic delay system supplies gas to the furnace only after air has been passing to the rooms for 10 to 15 minutes without satisfying the heating demand indicated by the thermostat setting.

University of Arizona Solar Energy Laboratory

The solar-heated laboratory at the University of Arizona was completed in 1959 and tested during the winter of 1959–1960 (1). The building is styled as a bungalow and used as a laboratory for solar energy research. It has 1,600 ft^2 of floor area, is well insulated, and has 435 ft^2 of double windows. The heat requirement at the design temperature of 30°F is 34,600 Btu/hr.

A unique feature of this system is the use of the solar collector not only as a heat receiver, but also as a heat radiator for cooling purposes. This 1,625-ft^2 unit covers the entire roof, and actually is the roofing as well as a heat exchanger. It is constructed of copper sheet with 120 integral parallel water tubes $\frac{5}{16}$" in diameter and 5" apart and connected at their ends to hot and cold manifolds. The material for the collector surface is shipped flat in long coils, 16" wide, and the tubes are inflated hydraulically at the construction site. These pieces are joined together on the roof by means of cleats to form a continuous metal surface. No glazing is used above the metal plate, and, for aesthetic reasons, a dark green paint has been used rather than black.

The heat-storage tank contains approximately 4,500 gallons of water. This tank is divided into two sections by means of a horizontal insulating baffle so that water at two different temperatures can be stored simultaneously.

Auxiliary heat is supplied by means of a small heat pump of 1½ hp. In a manner similar to that used in the Albuquerque building, hot water from the solar collector or from the hot section of the storage tank can be used to supply the heating requirements directly, when it is warm enough. At other times, it serves as the low-temperature heat supply to the heat pump, heat then being delivered to the building from the high-temperature side of the heat pump. Heat is transferred to the rooms by use of a continuous radiating ceiling constructed of the same type of material as that used on the roof. Water is circulated through 66 parallel tubes in 33 circuits. The heating panel completely covers the ceiling of all rooms, having a total area of 1,320 ft^2.

When operated as a cooler, water is pumped from the storage tank through the roof heat exchanger at night, radiating heat into the atmosphere. On a clear summer night, the large area of the unglazed unit permits discharge of up to ½ million Btu from water at about 70°F. On the following day, the cooled water is used directly in the ceiling panels for building cooling, or, if its temperature is too high, the heat pump chills it by transferring heat from it into the hotter, or condenser, section of the storage tank. This heat is dissipated the following night in the roof heat exchanger. Figure 3.4 gives a schematic diagram of this installation for winter heating and summer cooling.

FIGURE 3.4: SOLAR ENERGY LABORATORY—UNIVERSITY OF ARIZONA

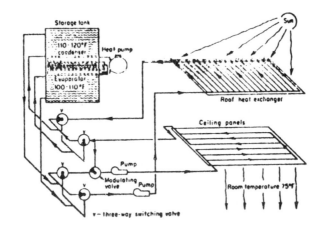

a.

Winter Heating System

(continued)

FIGURE 3.4: (continued)

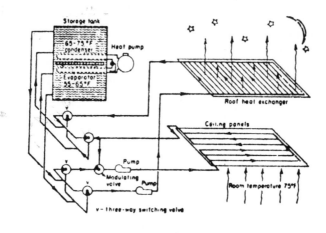

b.

Summer Cooling System

Source: AD-778 846, March 1974

Thomason Solar Homes

Thomason Solar Homes, Inc. of Washington, D.C., has for many years been designing and marketing solar-heated houses. The first of these has a floor area of
about 1,500 ft² and a solar collector of 840 ft² located on the roof and sloping
south wall. The heat-collecting surface is blackened corrugated sheet metal beneath one or two layers of plastic film and glass. Whenever the sun is shining,
water is pumped from a 1,600-gallon storage tank to the top of the collector
and allowed to run down on top of the corrugated surface to a gathering trough
and back to storage. Maximum water temperatures are in the 125° to 135°F
range. The water storage tank is surrounded by 50 tons of small rocks which
are intended to provide additional thermal storage by conduction and convection transfer from the tank.

The house is heated by circulating air around the tank and through the rock-
filled chamber to the rooms. When the available heat supply is not adequate,
the storage system is bypassed, and an auxiliary oil furnace supplies heat to the
air stream.

Cooling is also provided by use of day-night temperature difference. Daytime
removal of heat from the rooms is effected by circulating house air through the

previously cooled rock bin. Heat is then transferred to the tank of water, and the water is cooled at night by circulating it across a bare metal portion of the house roof (1).

The basic system is illustrated in Figure 3.5 for winter and summer use. This system has supplied most of the heat requirements despite half-cloudy weather and temperatures well below 0°C. In addition a substantial portion of the domestic water heating was achieved by solar heating. The system can store enough heat to last up to five cloudy, moderately cold days and it supplies 65 to 75% of year-round heating needs.

Several other systems have been developed and tested by the Thomason firm. Advantages and disadvantages have been determined for each system and pertinent components. Following are some minor alterations for later models.

A second Thomason solar house, built in 1960–1961, went through a number of changes. Cost for the original system was lowered, but the auxiliary heat cost ran slightly higher. An aluminum reflector was installed at the bottom of the solar heat collector to reflect additional sunlight onto the collector. The problem of summertime heat leakage from the solar heat collector into the closet space behind the collector was solved, keeping the closet cool.

A third solar home was improved in architectural appearance. A minimum of glass breakage was achieved using low-cost glazing. The heat collector was moved entirely up to the roof, permitting winter sunshine to enter the living room and a built-in swimming pool on the south side. In addition, improved air conditioning was installed.

A fourth house had a new type of solar heat collector using asphalt shingles and a new type of low-cost pancake heat storage. Both were used in an A-frame house. (See discussion of Thomason Sunny South House.)

A fifth house has been completed in Mexico. The house and the system were not constructed according to the designer's recommendation so the solar heating system does not provide the major part of the heat load. Although Mexico City is quite far south (19° N lat.), the temperature drops below freezing at times and can go into the low 20's.

Dover, Mass. House

Solar heating experiments were conducted for a few seasons after the completion of a solar-heated residence at Dover, Massachusetts, in 1949 (1). An air-heating collector in a vertical position on the south wall of the building supplied heat to storage bins containing cans of Glauber's salt, and no auxiliary heat source was provided in the design. The house has one floor of living space with an area of 1,456 ft^2. An attic is used for general storage and some of the experimental equipment.

FIGURE 3.5:   THOMASON SOLAR HOME—WASHINGTON, D.C.

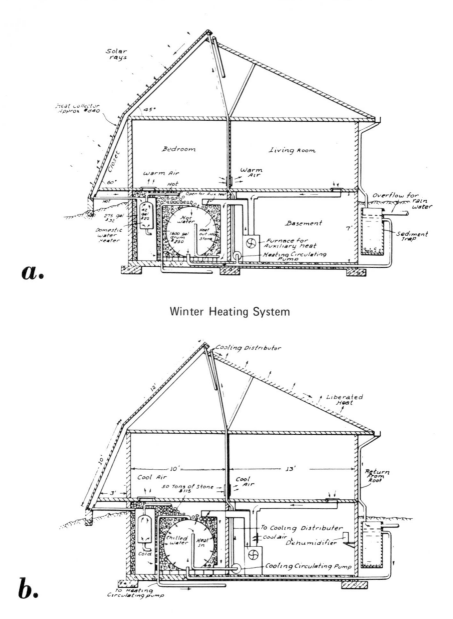

Winter Heating System

Summer Cooling System

Source: AD-778 846, March 1974

A 740-ft^2 solar collector occupied the entire south wall of the house above the ceiling level of the first story. Double glazing covered blackened metal sheets 4 by 10 ft in size, figured glass being used on the exterior surface. Air was circulated in a space between the metal sheets and an insulating back panel. Heated air was conducted by ducts to one of three storage closets located between various rooms of the house on the first-floor level. The heated air passed between stacks of 5-gal salt cans and, after its heat had been transferred to the salt containers, returned to the collector panels for reheating. The solar collector was divided into three sections, each connected directly to one of the storage closets (Figure 3.6).

FIGURE 3.6: SCHEMATIC OF HEATING SYSTEM—DOVER, MASS.

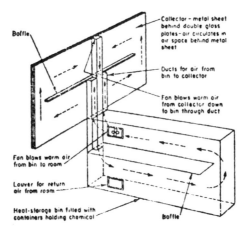

Source: AD-778 846, March 1974

State College, N.M. Solar-Heated House

A solar-heated house which has been publicized to only a limited extent was built in 1953 at State College, New Mexico. The house has approximately 1,100 ft^2 of heated living space in one-story and two-story portions. A 457-ft^2 solar air heater mounted at a 45° tilt is subdivided into three sections facing in a southerly and southwesterly direction. A blackened steel sheet covered with one layer of glass is used as the primary collector element.

Air is circulated from a storage unit through the collectors and back to the storage room, where 2 tons of Glauber's salt in 5-gal cans store the heat. Heat is transferred to the rooms by means of air circulated through the storage chamber and the necessary air ducts. Auxiliary heat is supplied by a gas furnace whenever the temperature of the storage room drops below 80°F.

Princeton, N.J. Solar Laboratory

A heating system employing solar-heated air, Glauber's salt thermal storage, and a vertical collector has been built and tested at Princeton, N.J. Typical January heat loads of about 12,000 Btu/hr were experienced in this 1,200-ft^2 laboratory styled as a modern house. Air was heated in a 600-ft^2 vertical collector in the south wall of the building by forced circulation across a black metal sheet overlaid with two glass covers. Approximately 275-ft^3 of Glauber's salt were used as storage. Collection efficiency of 46% was estimated for December and January, and the building was reported to have been solar-heated through two winters without auxiliary energy supply (1).

Japanese Houses

Of the three solar houses constructed in Japan in 1960, one, the solar space heating project at the Solar Research Laboratory of Nagoya, has been abandoned because of the lack of research staff. Two other solar houses, constructed in Tokyo and Funabashi, have operated successfully. All of these solar houses were heated by absorbed solar energy, but were cooled through a heat pump in such a way that the thermal energy was dissipated from the surface of a tubing sheet absorber during the night (4).

The collectors are of the water-heating type, with water conduits formed directly in thin sheet metal. There is no transparent covering over the collector, and the modest temperatures at which water is delivered from these units (12.8° to 26.7°C in winter) necessitate their being augmented by a 3-hp heat pump.

In one application described, a 2,460-ft^2 two-story dwelling is provided with a 1,410-ft^2 collector on a south roof tilted 15° from horizontal. Cold-side storage is in 9,600 gal of water in a basement concrete tank, and warmed water is stored in a 2,700-gal tank. Heat is transferred to the house rooms by circulating warm water through radiant ceiling panels. The efficiency of the solar collector on the house has been estimated to be about 22%, based on total incident radiation. About 70% of the heating demand was supplied by solar energy during three winters, or based on fuel equivalent of the electric energy used, about 42% of the load was carried by solar energy.

Capri, Italy Residence and Laboratory

A combination laboratory and residence in Capri, Italy, is heated by water circulated through 320 ft^2 of vertical, glass-covered metal radiator plates in the wall of a two-story, 1,940-ft^2 building. Hot water is stored in an insulated tank of 800-gal capacity and circulated by another pump to radiators in the rooms. Auxiliary heat can be supplied from a stove or electrical resistors (1).

Odeillo, France Space Heating Studies

In PB-223 536 J.D. Walton, Jr. of the Georgia Institute of Technology, which has been carrying out studies in conjunction with Centre National de la Recherche Scientifique (CNRS) under the direction of Professor F. Twombe, reports that the CNRS Solar Energy Laboratory has been active in the study and development of solar-heated houses since 1956.

The following sections describe general types of solar heating systems used for space heating at the Solar Energy Laboratory at Odeillo, France. Two of these systems involve some form of thermal storage and are used in single-family dwellings. The third system utilizes a thermal absorber only and is used to provide supplemental heat for the building which houses the Solar Energy Laboratory and the 1000 kw CNRS solar furnace.

Solar-Heated Houses: This house is shown in Figure 3.7. The south face of the house is covered with glass, behind which is located a concrete wall. The outside of the wall is painted black for maximum solar absorptivity and the concrete acts as a heat sink. Figure 3.8 is a schematic of the house. The objective of this design is to incorporate the absorbing surface, heat sink and warm-air circulation system into the structure of the house.

FIGURE 3.7: SOLAR-HEATED HOUSE—ODEILLO, FRANCE

a.

House Built in 1967

(continued)

FIGURE 3.7: (continued)

b.

Close-Up of South Wall

Source: PB-223 536, July 1973

As solar radiation passes through the glass **1** it is absorbed by the black coating **2** which heats the concrete wall **3**. Since the long wave radiation from the concrete is prevented from escaping through the glass, the air between the concrete and glass becomes heated and flows upward. As the hot air rises it goes through vents in the top of the wall and across the room. Simultaneously cool air is drawn in through ducts at the bottom of the wall, is heated and rises. This situation results in natural thermal convection and no other means of warm air circulation is employed.

FIGURE 3.8: SCHEMATIC OF SOLAR-HEATED HOUSE—ODEILLO, FRANCE

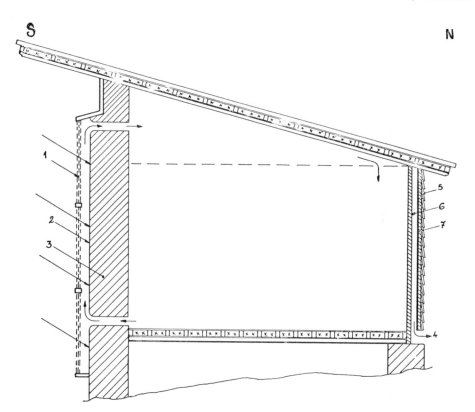

Source: PB-223 536, July 1973

Two of these houses were built in 1967 and have been occupied by engineers of the Solar Energy Laboratory since that time. Complete records were kept of construction costs and consumption of supplementary heat (electricity).

It was reported that the thermal energy provided by this system was about 600 kwh/m^2 per year (total solar collector amount, 80 m^2) and the cost was about \$40/m^2 to install. The total cost of the heat provided by this system was of the order of 0.8¢/kwh. During the past 5 years about ⅔ of the heat used in these houses was provided by solar energy. The volume of each house is about 300 m^3 and the insulation K factor is of the order of 0.9 to 1.0.

Figures 3.9a, 3.9b and 3.9c are of a model built by CNRS of a three-unit dwelling under construction. Figure 3.9a is a view of the model from the southwest while Figure 3.9b is from the southeast. Figure 3.9c shows the north side of the structure. The thermal system to be used in this dwelling was developed by researchers of the Solar Energy Laboratory.

FIGURE 3.9: CNRS MODEL OF THREE-UNIT DWELLING

View of South and West Walls

View of South and East Walls

(continued)

FIGURE 3.9: (continued)

C.

North Face

Source: PB-223 536, July 1973

Figures 3.10 and 3.11 are schematic representations of the thermal systems to be used in the three-unit dwelling. The major differences between this system and the one shown in Figure 3.8 are that windows are placed in the collector-heat sink concrete wall; the top air duct is divided into two smaller ducts; and a valve or damper is provided at the top duct which can direct the flow of heated air either into or out of the dwelling.

Figure 3.10 shows the dampers **8, 9** and **14** in the winter operating mode which directs the heated air into the dwelling. Figure 3.11 shows the dampers as they would operate in the summer. As the heated air is expelled to the outside through damper **8** cooler air is drawn into the house from the north side of the dwelling through damper **14**.

Figure 3.12 is a photograph of a solar-heated house constructed about 13 years ago. The solar collection area faces south and is covered by glass. As in the previous case, windows are provided in the collector area. This dwelling differs from the other two in that water serves as the heat sink rather than concrete. Therefore, the thermal collectors are essentially water radiators painted black to improve absorption. Water circulates through the collector and is stored in tanks in the area over the outside wall above the ceiling in the rooms. This dwelling is presently abandoned and construction costs and operational data were not obtained.

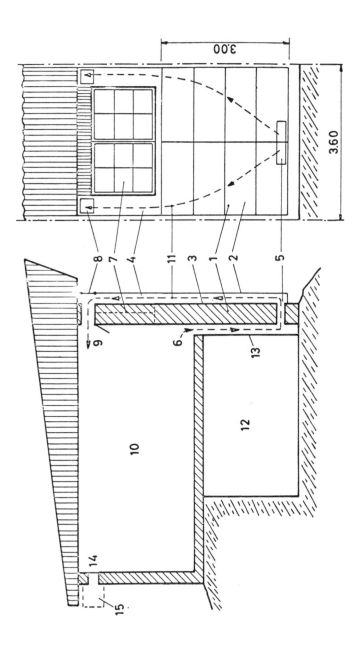

FIGURE 3.10: SCHEMATIC OF SOLAR HEATING SYSTEM OF THREE-UNIT DWELLING—WINTER OPERATION

Source: PB-223 536, July 1973

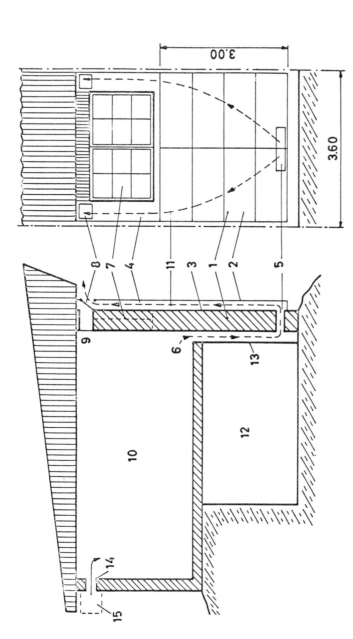

FIGURE 3.11: SCHEMATIC OF SOLAR HEATING SYSTEM OF THREE-UNIT DWELLING–SUMMER OPERATION

FIGURE 3.12: SOLAR-HEATED HOUSE BUILT ABOUT 1962

Source: PB-223 536, July 1973

Solar-Heated Building: The nine-story building which houses the Solar Energy
Laboratory and the parabolic reflector for the 1000-kw solar furnace receives
about one-half of its heating requirements from solar energy. In this case black
corrugated metal panels are strategically located behind the glass panels which
cover the east, south and west walls of the building. A schematic showing the
principle of operation of this solar heating system is shown in Figure 3.13.
Solar radiation enters through the glass **1** and heats the corrugated metal panel
2 which is contained in the area defined by the glass and duct **3**. As air between
the absorber panel and glass is heated it flows upward and out through the
vent **4** while cooler air is drawn down between the back of the absorber panel
and the duct.

As pointed out previously, no provision is made for storing heat in this system
and it is effective only when the sun is shining. Nevertheless, the temperature
in the offices and laboratories is maintained at a relatively constant level and
has been very comfortable during all of the time that Georgia Tech has occupied
space in the building. Even during February, auxiliary heat was required only
at night and on overcast days.

Future Activities: In addition to the original three-unit dwelling described previously, construction has begun on 31 additional units. All 34 units will be located near the 1,000-kw solar furnace. These houses will use the current state

of technology related to the use of solar energy for space heating. Some units will use water for heat storage. Various building materials and types of insulation will be evaluated in order to optimize various structures with respect to efficiency, economics, size of dwelling, collector area, etc.

FIGURE 3.13: SOLAR HEATING SCHEMATIC FOR CNRS BUILDING

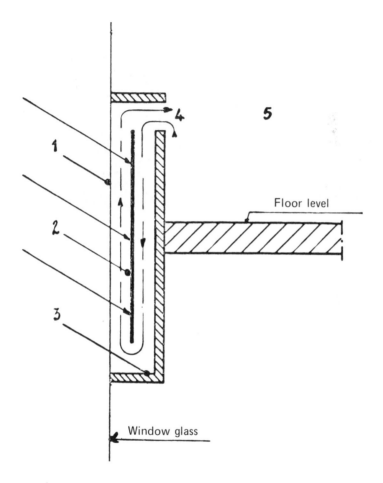

Source: PB-223 536, July 1973

The Georgia Institute of Technology Engineering Experiment Station plans to arrange with the researchers of the Solar Energy Laboratory to collect and evaluate construction costs, efficiencies, economics and overall capabilities of the dwellings being constructed at Odeillo.

Delaware House

A sophisticated version of solar utilization is an experimental house, the Solar One, designed by the Institute of Energy Conversion at the University of Delaware, Newark, which was dedicated in July of 1973 (Figure 3.14).

FIGURE 3.14: DELAWARE HOUSE

Source: AD-778 846, March 1974

In the 1,350-ft², two-bedroom dwelling, described by Institute Director Dr. Karl W. Boër, flat-plate solar collectors containing thin-layer cadmium sulfide-copper sulfide solar cells will convert 50% of the incoming solar energy into heat and 6% into electrical energy (3). 24 plastic-covered skylights on the 45°-angled roof contain panels of cadmium sulfide-copper sulfide photovoltaic cells, which will eventually produce 20 kwh a day of electricity, and solar heat collectors. 6 additional solar heat collectors a on the house's south side supplement the roof collectors.

According to Dr. Boër, (PB-223 536), to reduce deployment cost and to provide a leak-proof roof, the collectors are deployed between the roof joists from the inside; the outside is glazed with ¼" plexiglas with support posts to avoid excessive sagging providing a skylight enclosure. The size of the collectors is 4 x 8 ft² and determined by minimizing shading edge effects and connector cost (wide collectors) as well as minimizing glazing and roofing costs (narrow collectors). The collectors are not exposed to weather, hence can be fabricated inexpensively using, e.g., corrugated paper, with fiber glass enforcement, in situ foaming, thin anodized aluminum foil for mounting of the CdS solar cells and a thin glass sheet as a seal against humid air. A cross section through such a collector is given in Figure 3.15.

FIGURE 3.15: CROSS SECTION OF COLLECTOR CONTAINING SOLAR CELLS

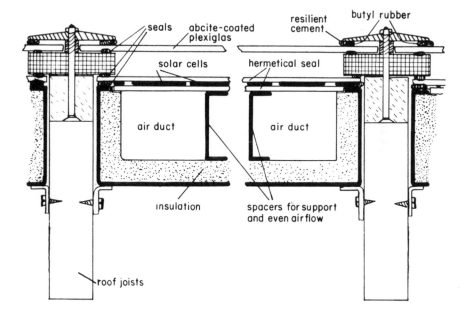

Source: PB-223 536, July 1973

Heat transport fluid is air to reduce collector cost, increase its lifetime and its reliability (avoid liquid leakage). The heat is transferred only from the back through 2" deep air channels to avoid dust deposition on the front surface which could necessitate maintenance.

Spray nozzles near the top edge of the roof are proposed to release a small amount of detergent during heavy rain, to break down greasy dust deposits and to provide automatic clearing of the outer roof surface. A cost efficiency analysis coordinated with design improvement is planned to develop optimized collector design. Special attention will be given to the use of inexpensive materials, ease of deployment, increase of collector life, reduction of collector losses and collector esthetics.

In the Delaware house, solar radiation heats the solar cells at the same time that it produces electrical power. In the winter, air from the basement is pumped into the roof area containing the hot solar panels, then recirculated into the basement where the heat melts a salt or a eutectic salt mixture encased in a 6 x 6 x 6 ft plastic container. During the evening hours when the house begins to cool down, the cooler air is circulated through the salt, extracting its heat of fusion and heating the air. During the summer, the heat pump is used as an air conditioner, operating during the evening hours to freeze another eutectic salt mixture. Then, during the day, warm air is circulated through the cold mixture, cooling the rooms.

The Delaware house uses sodium thiosulfate during the winter cycle and a mixture of sodium chloride, sodium sulfate, ammonium chloride, and borax during the summer cooling cycle. There are some difficulties in getting the salts to melt congruently and in some cases the molten salts must be nucleated before they will recrystallize. However, these problems are not insurmountable.

The thermal system makes extensive use of heat storage in conjunction with a heat pump to maximize the use of solar energy and to shift the operating time of the heat pump into off-peak hours (night). Figure 3.16 shows a simplified cross-section elevation schematic of the solar house to indicate the major features of its operation.

During the heating season the harvested heat will be stored in a reservoir using $120°F$ heat of fusion ($\sim$ 1 million Btu maximum storage). When the collector temperature drops below $130°F$, an alternative duct will be opened to store heat in a $75°F$ base reservoir ($\sim$200,000 Btu capacity). From here the heat pump operates to charge the $120°F$ reservoir (during night hours) or to heat the rooms directly. This may be done even after the base reservoir is depleted and if the collection temperature drops below $75°F$ but is still in a range at which the use of a heat pump is efficient (e.g., at an inlet temperature above $50°F$).

During the cooling season, night radiation is used to freeze the $75°F$ salt eutectic and to cool the base exchanger of the heat pump. Coolness produced by the heat pump during night hours is stored in the $55°F$ eutectic interspaced with the $120°F$ salt containers within the main heat reservoir.

Hot water is produced via air heating using an aluminum fin air-conditioning-type heat exchanger placed in the main air duct. A water tank will be used with proper tube connections to aid stratification and with an electric booster at the outlet pipe.

The solar system has fail-safe vents with weight-loaded louvers to prevent overheating by using chimney-action air cooling. The main purpose of this solar house is to show feasibility for a major contribution of solar energy to the energy balance of the house, and to shift its main use of auxiliary electric energy into off-peak hours. The results obtained from this house shall provide the basis for developing a prototype house of attractive cost efficiency, and to demonstrate incentives to power utilities to utilize distributed solar conversion systems.

FIGURE 3.16: CROSS-SECTION ELEVATION SCHEMATIC OF THE SOLAR HOUSE

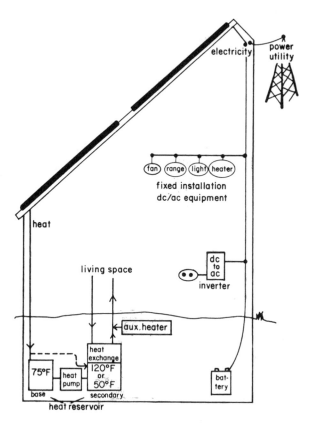

Source: PB-223 536, July 1973

Zomeworks House—Drum Wall and Skylid

According to S. Baer (PB-223 536) a drum wall has been built into a house in Albuquerque N.M. The particular drum wall is a south wall made up of 55 gallon drums or other containers filled with water. The drums are stacked immediately behind a south window so that sun shines through the window onto the drums, heating them and the water within them. A door, or other insulating means is closed over the south window during the nighttime to lessen the heat loss to the outside.

The drums form the south wall of the room and give heat to the room by radiation, conduction and convection. A curtain can be placed between the room and the drums to control the flow of heat from the drums to the room. Another means to control the flow of heat from the drums is to build an inner insulating wall with vents at top and bottom, thus restricting the transfer to convection but allowing great control.

The drums are effective at low temperatures to below 90°F. The heat transfer from the metal or plastic container walls to the water is excellent. These two characteristics combine to make the collection efficiency superior to almost all other heat collectors. On clear days in December the drum wall of the Baer house is calculated to have collected 1,430 Btu/ft^2, ±10%.

55-gallon drums are inexpensive. The drums can be stacked upright with 1" thick lumber sandwiched between rows. Or they can be placed on their sides in supporting racks. Glazing can be done with single strength window glass siliconed to the drum racks or to the wood spaces between rows of drums. The insulating door can be of sandwich panel construction with a reflective interior skin which reflects sunlight into the collector when it is open. A drum wall can be installed for less than $5.00/ft^2.

A drum wall is best used for solar tempering in conjunction with supplementary heat such as a wood stove or gas burner. In a house where the temperature can fluctuate 10° to 15° without great discomfort a drum wall can effectively do 75% of the space heating. Rooms not adjacent to the wall must draw their heat by airflow from a room with a drum wall.

A drum wall can combine as a summer cooling system by opening the door during the night or even by just opening windows at night and allowing cool air to chill the drums. South-facing drum walls are effective at latitudes from 30° to 45°F.

Skylid: The skylid is an insulating cover that fits beneath a skylight which can open to allow sunlight to enter or to close to prevent heat loss. The skylid is powered by two containers connected by a tube—one on each side of the insulation. Enough Freon is introduced into the system to fill one container. The cover is pivoted and balanced so that as the Freon moves from one container to the other the cover opens or closes. Temperature differentials then open the cover when the sun comes up and close it when it goes down. The skylid is shown in Figure 3.17 as it would appear installed in a residential interior.

Bead Wall: This bead wall is a method to change a clear window with a U factor of 0.5 into an opaque wall with a U factor of 0.1 or less in a few moments. The bead wall operates by a pneumatic system of filling and emptying a cavity between two glass panes with specially treated plastic foam beads.

Applications of the process are expected wherever people have previously used curtains or blinds and also as part of solar heating systems (such as in conjunction with the previously mentioned drum walls). The method is expected to make houses constructed entirely of double glass panels sensible for people or plants.

FIGURE 3.17: ZOMEWORKS SKYLID

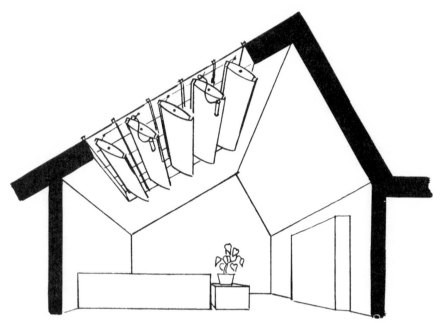

Source: Zomeworks Corporation, Albuquerque, N.M.

Sky Therm Houses, Phoenix and Atascadero

A simple, passive, relatively inexpensive solar heating and cooling system for
buildings with the trade name Sky Therm has been developed by H.R. Hay of
Sky Therm Processes and Engineering, Los Angeles. Two model prototype residences have been built in Phoenix and Atascadero, California; the latter is shown
in Figure 3.18 as described in PB-235 432.

The system utilizes confined water ponds in intimate contact with the roof or
south wall elements of the building, and movable panels of rigid insulation acting as a thermal valve in controlling heat flow into and out of the water ponds.
In winter, solar energy is allowed to heat the water pools, hence the building
structure, during the day. At night, heat loss from the building is prevented by
placing the insulation panels over the water ponds. In summer, the insulation
panel movement is reversed; during the day, the insulation is placed over the
water pools restricting heat gain, then removed at night to allow heat loss to
the cooler environment.

Sky Therm is not limited to the one-story, flat-roof style shown; designs are
under development for south wall heating in northern and snow regions.

FIGURE 3.18: SKY THERM HOUSE

Source: NSF-RA-N-73-004, March 1973

Sky Therm provides new, more comfortable standards by mildly heating and cooling through an acoustical metal ceiling—no hot or cold spots—no air drafts or noise. Regular building materials are the heating and cooling system; hence low initial cost compared with conventional design.

Phoenix, Arizona, tests established that Sky Therm Natural Air Conditioning required no conventional heating or cooling when air temperatures ranged from subfreezing to 115°F. This result is unequaled by any other system for solar energy use.

A ¼-hp motor running only two minutes morning and night (or manual operation) moves the insulation panels (thermal valves) on trackways from over water beds to a carport or patio area for 3-deep stacking. In some climates, a little more electricity may be needed.

A schematic cross section of a typical Sky Therm system installation is shown in Figure 3.19. A rooftop location is depicted but a south wall installation utilizes the same type elements and operational features, with differences in configuration and dimensional details only.

The system is supported by an interlocking, corrugated, galvanized roof deck, having a three-inch deep cross section, which is also the ceiling of the building interior. The system is best integrated with a building constructed of concrete block, filled with high heat capacity material, for both exterior and interior walls. This type of construction provides the structural support for weight of the water ponds while acting as a thermal fly-wheel in leveling out high and low temperature swings within the building.

Above the roof deck, 2" by 7" steel box beams are mounted to support the movable insulation panels. Aluminum trackways, mounted on the beams, provide for horizontal panel movement and prevent panel lifting due to wind loads.

The insulation is 2" thick sheets of clad, closed-cell polyurethane which are mounted in channels and H members to form large panels. The panels ride on wheels within the trackway and are moved by a motor-operated chain and sprocket arrangement.

FIGURE 3.19: SKY THERM SOLAR SYSTEM SCHEMATIC

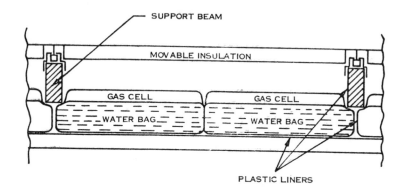

Source: PB-235 432, May 1974

The water beds are fabricated of either laminated polyethylene or PVC plastic. With the former material, two 2-mil thick films are laminated to form a liner which covers the roof deck and is lapped across the beams to form a sealed roof over the building. The liner is covered by a low-cost four-mil black polyethylene sacrificial liner that prevents ultraviolet degradation of the underlying liner. The water bags are formed of 4-mil clear polyethylene constructed in tubular shape, folded up and fastened at the ends above water level.

The PVC water beds are constructed from a bottom sheet of pigmented swimming pool liner, 20 mils thick, sealed to two clear ultraviolet-resistant sheets. The pigmented liner, with side extensions lapped across the beams is sealed to one layer of the clear vinyl to form a water bag. The second clear layer is sealed to the first clear layer to form a gas well over the water bag. The gas cell retards the loss of heat from the water bag to ambient air.

The Sky Therm system is essentially passive in the sense that static water beds collect and store solar energy, and transfer it naturally by conduction to the building living space. In contrast, competing solar heating and cooling systems use fluid circulation between solar collectors, thermal storage units and space air heat exchangers to achieve the required energy transfer. There is one action, the movement of the insulation panels, that uses an electric motor, but this operation can be performed manually if necessary; thus, the Sky Therm system is not dependent on auxiliary power for its operation.

The positioning of the insulation panels is controlled by a manually set timer that opens and closes the panels at the desired times in both the heating and the cooling seasons. A parallel control circuit allows manual control of the chain and motor gear to provide for abnormal operation of panels.

Modifications to the basic operating mode of the Sky Therm system are possible for unusual conditions. During hot spells the water bags can be covered with a water layer to obtain additional cooling by the nocturnal evaporation of the water. Circulating air fans may be used within the building to lessen vertical temperature gradients and hence, increase comfort. It is also possible to circulate heated (or cooled) water from the water bag to fan coil units throughout the building. This latter option can be used in conjunction with the Sky Therm (as it is in other solar-type systems) as a supplementary conventional heating and cooling system in climatic zones where building energy requirements cannot be totally met by the Sky Therm system.

Thomason Sunny South House

Another Thomason house is a design (Figures 3.20 and 3.21) using a rooftop pond from which the solar warmed water will flow by gravity to the pancake heat storage under the floor. The floor is thus warmed and the rising heat warms the living quarters above. The water will be pumped to the pond (only 8 to 9 ft above the heat storage tank) within 30 to 60 minutes at a 4 lb/in^2 pressure.

The solar rays (Figure 3.21) come in at a low angle, 30° to 45°, and are spread out over the entire large horizontal pond area. The solar flux is only about 50 to 70% as intense as it would be if the sunlight were shining directly on a surface normal to the sunrays. This means that the rays are deconcentrated and the pond must therefore be 50 to 100% larger to intercept a given amount of solar heat. Such a large pond loses substantial heat upwards during the day through evaporation and wind convection; the pond also loses some heat at night through the large insulated area, which cannot be completely effective.

To minimize the drawback, the designers reduced the size of the rooftop pond by introducing a reflector R adjacent to the north edge of pond. Thus, the solar rays B (Figure 3.21) that would normally bypass the pond, are reflected onto the pond. The added reflector, with a height approximately 60% of the pond's width, will intercept as much solar energy as the pond itself. If a reflectance value of 70% from the reflector is assumed, then the increase of solar energy input to the pond would be approximately 170% of that which would strike a level pond directly. However, some of the morning and afternoon rays are not reflected directly into the pond, so that the overall energy input between 9:00 AM and 3:00 PM would be less than 170%.

From late February through March the sun is about 40° above the horizon in Washington, D.C. (about 45° in Phoenix). In designing for a maximum reflector-to-pond ratio for that period the reflector needs to be approximately as high as the pond is wide, as illustrated in Figure 3.21 (for example, pond 10 ft, reflector 10 ft). However, the ratio and angle are not deemed to be critical, and some deviation may be desirable for architectural purposes or other considerations.

FIGURE 3.20: THOMASON SUNNY SOUTH MODEL

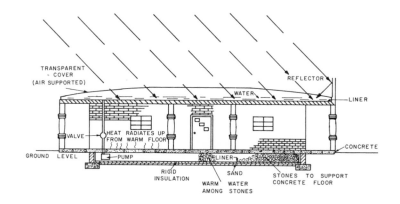

Source: PB-223 536, July 1973

Figure 3.22 shows a design for a large flat-roof sturcture which requires considerable architectural deviations. Ponds **1, 2, 3**, etc., are fairly narrow in the north-south direction. Relatively short reflectors R^1, R^2, etc., are adjacent to their respective ponds. The ponds may extend the full length of the building in the east-west direction (for example 30 to 70 ft for a home or perhaps several hundreds of feet long for a commercial building). Longer ponds and longer reflectors will have a greater percentage of the reflected light striking the pond during morning and afternoon periods.

Much heat can be liberated from a rooftop at night. This is especially true if the relative humidity is low, there is a good breeze, and the nighttime temperature drops to about 60°F or below, or if there is no cloud cover to block radiation to the night sky.

Employing that principle in the rooftop pond system the water will be pumped to the roof at night. It may flood the transparent cover, or the transparent cover may be removed for the summer thereby permitting ready evaporation, radiation and conduction to the cool night air. Experiments will be run to determine whether it is best to allow all of the water to flow back to storage each morning to cool the home, or to leave a portion of the water in the rooftop pond to absorb incoming solar energy and reduce ceiling heating. The answer to this question may be different under differing conditions.

FIGURE 3.21: ROOFTOP POND

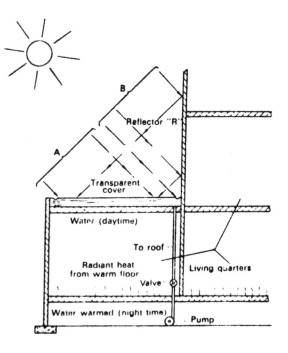

Source: AD-778 846, March 1974

FIGURE 3.22: SERIES OF ROOFTOP PONDS

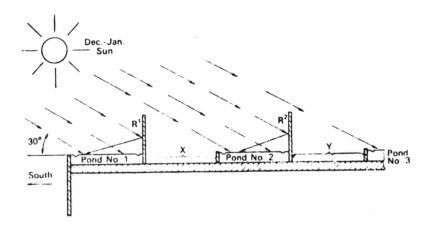

Source: AD-778 846, March 1974

For example, during mild weather in autumn and spring the water likely can be left in the pond day and night. During the hot summer it is likely that some of the cooled water should be returned to storage to help keep the house cool. To assist in keeping the house cool, a blower may be used to circulate the air, or a simple fan may be used to blow the air against the cooled floor and circulate it through the home (2).

Shiraz, Iran Building

The Mechanical Engineering Department of Pahlavi University, Shiraz, Iran has designed, built and tested a solar-heated building as part of the Mental Hospital, 10 miles east of Shiraz (Figure 3.23).

FIGURE 3.23: SOLAR-HEATED BUILDING—SHIRAZ, IRAN

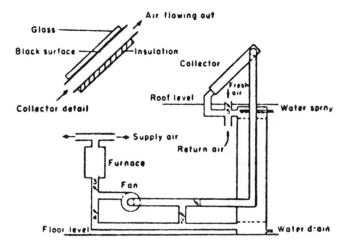

Source: AD-778 846, March 1974

It is a one-story building with 5,000-ft^2 floor area, accommodating 20 to 30 patients. A warm-air system is presently employed for heating in winter, with no summer cooling provided. For this experiment, a flat-plate solar heat exchanger for collection of solar energy using air as the transport medium and rocks for solar energy storage, was selected.

Air heated in the collector to the desired temperature is either sent directly to the rooms or to the rock bed, depending on the building heat demand. The returned air from the building is circulated back to the collector. During nights and cloudy days the air is circulated through the storage in the opposite direction. No attempt is made to utilize warm air for water heating. A separate solar water heater may be employed for this purpose. The single-glass, flat-plate solar collector has an area of 10 ft^2 and an 8-ft^2 rock storage and is designed for solar heating and economic evaluation.

Based on this experiment, it has been found profitable to use solar heating in Iran, which features low annual precipitation and relatively cold winters (5).

U.S.S.R. System

In the U.S.S.R. the Physicotechnical Institute of the Uzbek Academy of Sciences jointly with the Tashkent Regional Scientific Research Institute for Experimental and Standardized Planning conducted experimental research during 1968–1970 on a combined operation of heat pump and solar installations for space heating and air conditioning. The experimental chamber has an area of 6.5 x 4.5 m^2 and a height of 2.7 m. During tests, adjacent rooms were heated by a conventional boiler room radiator system (Figure 3.24).

FIGURE 3.24: HEAT PUMP AND SOLAR HEATING ARRANGEMENT
U.S.S.R.

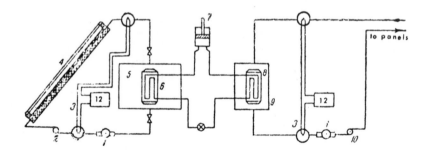

Source: AD-778 846, March 1974

During clear or partly cloudy days in winter, a 3-m^3 capacity heat reservoir **1** containing condenser **2** and heat pump **3** was activated. The solar collector **4** is made of stamped steel plate and placed in an insulated wooden box covered by double glass plates. Two such wooden boxes, each 6 x 1.25 m^2, are placed at 50° elevation angle. The energy output was measured by a water pump **5** and differential thermocouple **6** equipped with a potentiometer **12**.

The heat pump installation was served by a small FAK-0.7 E cooler unit having a pressurized type 2VF 4/45 compressor and an 0.6 kw motor; Freon-12 was used as the working medium. The thermal capacity of this heating system for economical consideration must be calculated at 65 to 70% of maximum demand. Based on several years actinometric data, heat storage capacity should be designed to allow for 3 to 4 consecutive cloudy days; the occurrence probability of longer overcast intervals is less than 20%.

The heat pump condenser **8** was installed in a small reservoir **7**. Input of heat and cold consumed by the test chamber is measured by a water meter **5** and a differential thermocouple **6**. To avoid clogging of heating and cooling coil pipes, the reservoir is fed with chemically treated water. The closed cycle flow of cold-hot media inside the reservoir is done by a pump **9**, while water circulation through the solar heating unit is achieved by another pump **10**.

Experimental tests have shown that the combined operation of solar heating and standard heat pump and radiator system makes it possible to use comparatively low temperatures of the heating-cooling agent, which increases the conversion coefficient of the heat pump and efficiency of the solar heating unit (6).

SOLAR COOLING AND REFRIGERATION SYSTEMS

The following is a brief review of solar cooling activities in various countries derived from AD-778 846.

Australia

In Australia much effort has been expended to produce cooling by other natural processes without resorting to expensive refrigeration. Natural air conditioning is particularly pertinent here because the continent lies mainly between $10°$ and $35°$ south latitude, and is characterized by vast desert areas where extremely high temperatures and low humidities prevail.

Experiments with rock piles and plastic rotary generators have been conducted, and a new development of a unit air cooler using a plastic heat exchanger with evaporation-cooled plates is in the testing stage. The operating principle of this cooler is simple. Hot outdoor air is blown through passages formed by dimpled heat exchanger plates with every alternate passage traversed by room air into which water is sprayed. The room air is cooled by the evaporative process, but its humidity is increased at the same time and so this air is exhausted to the atmosphere after it has completed its cooling function of the heat exchanger plates.

The basic unit which has been tested is a cooler suitable for a single room, but it is thought that it can be extended to meet the needs of a typical dwelling. Since the cooling process is one of essentially constant moisture content, the process is not suitable for regions with high humidities, but it does hold promise for large areas in Australia where extremely high temperatures are prevailing with moderate or even low absolute humidity (10).

Brazil

Since 1960, the Center for Studies in Applied Mechanics (CEMA, organized in 1952 as an affiliate of the National Institute of Technology, Rio de Janeiro) has conducted a study of problems on the internal arrangements of refrigerating units and harnessing of solar energy by means of a fixed conical reflector specially designed for applications of this type.

Figure 3.25 gives a diagram of the refrigerator, based on a generator consisting of the heating chamber **1**, generator casing **2**, vapor outlet duct **3**, and separator **4**. A two-stage condenser is used, the first stage **5** being outside the cold chamber **14** and the second stage **7** inside it; the two stages are separated by check valve **6**. The evaporator **10** is preceded by the liquid outlet capillary duct **8** and the expander **9**. The absorber **11** is connected to the generator by constant-pressure valves **12**, with an auxiliary coil equipped with control valve **13**. The cold chamber forms the interior of the refrigerator **15**.

FIGURE 3.25: SOLAR REFRIGERATOR UNIT—BRAZIL

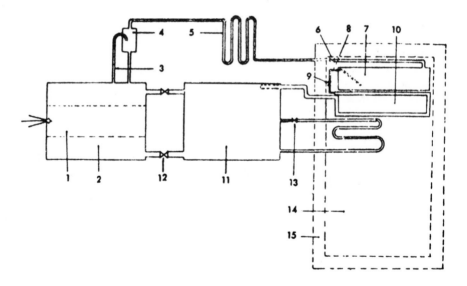

Source: AD-778 846, March 1974

The generator **2** which contains a rich 40% solution of ammonia in water, receives the solar heat concentrated either inside it (when a solar cell or a parabolic concentrator is used), or outside the cylindrical receiver (when a conical reflector is used). The NH_3 vapor liberated from the ammonia solution by heating it to $70°C$ under 8 atm pressure passes through duct **3** into the separator, which frees the ammonia vapor from the entrained water vapor. The water vapor content is relatively low, 1 to 5% (according to the temperature), but it is important that it be removed to prevent icing of the ducts.

From the separator, the vapor passes into the coil 5, where its temperature falls to about 50°C with a corresponding pressure drop to 5.2 atm. The second stage of the condenser is inside the cold chamber, so that part of the excess cold is utilized for condensation, thus eliminating moving parts, such as circulating pump or fan. From the liquid ammonia tank 7 the capillary duct 8 feeds the expander 9, which discharges into the evaporator 10 where the pressure is held at about 3.6 atm.

In the absorber 11, where the pressure is only 3.0 atm the vapor is continuously absorbed, progressively enriching the solution. The auxiliary coil 13 holds the pressure constant until the pressure in the generator falls below 3 atm. As soon as this level is reached in the generator, NH_3 will be exchanged through the constant-pressure valves 12, regenerating the rich solution, so that the cycle can be repeated.

The required heat will be supplied during 4 hours of exposure to sunshine, and the capacity of the condenser 7 is sufficient to ensure uninterrupted operation for 24 hours. The refrigerator box has a useful capacity of 9 ft^3 (about 255 liters) or about 12 ft^3 of gross internal volume. The prototype unit was designed for 32 kg of NH_3 (16 kg of which were effectively circulated) at 48 kg of water, or a total of 80-kg rich solution in the generator and 64 kg of diluted solution in the absorber at the beginning of the cycle. The condenser charge is sufficient to maintain the cold for 24 hours (7).

France

The Mont Louis Solar Energy Laboratory has developed a refrigerator operating on an intermittent ammonia cycle, with the ammonia solution directly heated in the zone of concentrated solar radiation. The ammonia solution is heated during insolation, yielding ammonia gas under pressure, which is then liquified. The cold (due to the distillation and redissolving of the ammonia gas) is generated after sunset and during the night.

The apparatus, made entirely of welded steel, comprises the following parts (Figure 3.26). A reservoir tank A is completely filled with ammonia solution; the tank must remain cold. During the various phases of operation, its upper part contains solutions richer in ammonia than the solution in its lower part. Heat exchanger B consists of concentric pipes. Solar energy is concentrated on heating tube C, by a cylinder-parabolic mirror made of aluminum-magnesium alloy, heightened by anodizing. The focal length of the mirror is 25 cm. It rotates about its optical axis, which has an east-west orientation, by means of circular metal supports, so as to obtain the inclination giving optimum concentration of energy on tube C.

FIGURE 3.26: REFRIGERATION UNIT—FRANCE

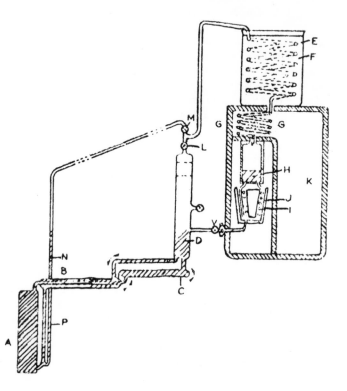

Source: AD-778 846, March 1974

The boiler **D** contains enough liquid to permit the distillation of the quantity
of ammonia gas that can be produced in a single day. Condenser **E**, cooled by
a nonreplenishable water reserve **F**, is placed in a tank protected against solar
radiation by a suitable coating. A liquid ammonia reservoir **H** is surmounted
by coil **G** and connects with another coil **I** surrounding the ice freezer **J**; the
entire assembly is contained in a cold chamber **K**. Valves **L** and **M** permit
distillation of the ammonia during the day and its reabsorption at night in the
ammonia solution, which has become diluted during the day. The outlet **V**
serves to clean the liquid residues from the evaporator **I** at the end of a refrigerating period.

Tests have been conducted on one apparatus of 1.5-m^2 solar collecting surface,
and a smaller one with 0.18 m^2 area. The daily output of ice was 95 and 6 kg
respectively, with the actual duration of ammonia gas distillation ranging from
4.5 to 5 hours, not counting the preheat period of about 1.5 hours (8).

India and Pakistan

A preliminary technical and economic feasibility study was conducted in the 1960s to establish the possibilities of using solar energy for space cooling in India and Pakistan. The system under consideration (Figure 3.27) consists essentially of a flat-plate solar heat collector coupled to a specially designed absorption refrigeration system. The absorption refrigeration unit used in this system is designed to operate at a lower generator temperature and at a relatively higher evaporator temperature ($55°F$) than in conventional systems, giving a very high effectiveness of heat and mass transfer throughout the installation.

**FIGURE 3.27: DIAGRAM OF SOLAR HEATING AND AIR CONDITIONING
SYSTEM**

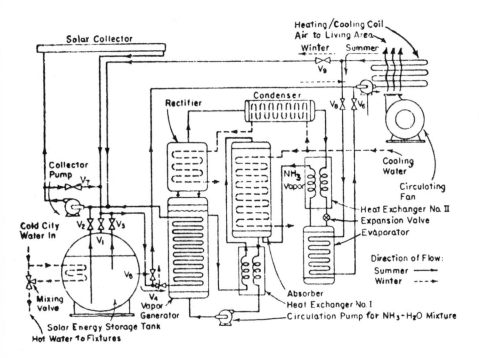

Source: AD-778 846, March 1974

Tests on this design have been conducted at MIT, using a flat-plate collector similar to the one in the MIT Solar House IV. The air conditioning system essentially consists of a solar heat collector, an ammonia absorption cooling unit, an air-to-water heat exchanger, and a heat storage tank. The main components of the cooling unit are a generator, rectifier, condenser, evaporator, water cooling coil, and an absorber.

It was determined that such a system would be technically feasible in a climate such as that of New Delhi. Most of the cooling load occurs during the periods of high solar incidence and hence very little overnight storage is necessary. For the month of May the peak value of available refrigeration is 84.5 Btu/ft^2 per collector-hour. This system would become economically attractive if a solar heat collector could be built to last long enough to be depreciated over 10 years, under India's and Pakistan's climatic conditions (9).

U.S.A.

Phoenix Sky Therm House: In Phoenix, Arizona, a prototype building has been kept within the comfort zone during a period of more than 18 months by the operation of a unique solar heating and sky-cooling system as described previously. The structure uses shallow ponds of water which are in thermal contact with the metal ceiling of the room to provide both thermal storage and temperature modulation. Horizontal plastic panels above the ponds constitute the roof of the building, and these can be pulled away from the ponds during winter days to permit the rays of the sun to warm the ponds and thus heat the house. In the summer the situation is reversed and the insulating panels are removed at night so that the ponds can be cooled by evaporation and by radiation to the sky (10).

University of Florida Study: The Solar Energy and Energy Conversion Laboratory of the University of Florida conducted tests on a jet refrigeration system using flat-plate or nonconcentrating collectors which functioned even on cloudy days (11). The laboratory has been experimenting both with small and large (5 ton) units. Flat-plate collectors heat water which is then circulated to drive the ammonia from the water in the system's generator.

The ammonia vapor is condensed and then expanded to provide evaporative cooling; the ammonia vapor is then reabsorbed into the water to repeat the cycle. The solar absorber in some systems is combined with the ammonia generator and all the equipment is installed behind or under the absorber. A small 4 x 4 ft unit can produce 80 lb of ice on a good day.

It should be emphasized that the applications mentioned for this design do not require concentration of solar energy, and therefore use both the direct and the diffuse portion of solar energy (11).

American-Saint Gobain and University of Wisconsin Study: Experimental programs directed toward the use of solar energy for absorption air conditioning have been carried out since 1963 by the American-Saint Gobain Corporation and the University of Wisconsin. In the American-Saint Gobain studies, steam at atmospheric pressure was produced in a heat exchanger through which solar-heated air was circulated.

In practical application, the steam would be used as the heat source in a commercial type of absorption air conditioner formerly manufactured by the

Servel Corporation. The refrigerant generator in one model of the Servel air conditioner was heated by steam produced in a small gas-fired boiler, for which the solar collector and heat exchanger would be substituted. The object of this investigation was therefore the use of an existing conventional heat-operated air conditioning system in so far as possible, with the conventional heat source being replaced or augmented by solar energy.

The Servel air conditioner, now being manufactured by the Arkansas-Louisiana Gas Company (Arkla), employs an absorption-refrigeration cycle with water vapor as refrigerant and lithium bromide solution as absorbent. A schematic diagram of such a system is shown in Figure 3.28. The cooling effect is produced in the evaporator, where water vaporizes at low pressure and at temperatures in the $40°$ to $50°F$ range. The vapor is then absorbed, and the resulting lithium bromide solution is pumped to the generator, where heat from burning gases produces water vapor for subsequent condensation and return to the evaporator.

FIGURE 3.28: ABSORPTION-REFRIGERATION CYCLE SYSTEM

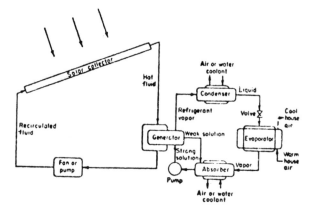

Source: AD-778 846, March 1974

In the experimental program directed toward the design of a suitable combination of this unit with a solar heat supply, a solar air heater of 128 ft^2 at a $27°$ slope with the horizontal ($40°$ latitude) was used. Various glazing arrangements were tested, and heat-recovery efficiencies at temperatures necessary for steam generation were measured. Since the data indicate that hot air can be supplied from an overlapped-plate type of solar collector to a refrigerant generator at temperatures above $250°F$ at efficiencies above 25%, the operation of a unit of the Servel type by means of solar energy is entirely possible.

Other media can be used for supply of solar heat to a refrigerant generator. Water from a solar hot-water heater at a temperature well below the boiling point would be a satisfactory heat source for a refrigerant generator operating at 170° to 180°F. Other arrangements could involve steam supply from a collector similar to the solar water heater, and the refrigerant generator itself could be a large solar collector (1).

U.S.S.R.

The Physicotechnical Institute of the Turkmen Academy of Sciences has designed and tested a solar absorption refrigerating unit with an open-type regenerator. This installation will be used for space air conditioning in dry sunny southern regions of the Soviet Union. The system (figure 3.29) works in the following way.

FIGURE 3.29: SOLAR ABSORPTION REFRIGERATING UNIT—U.S.S.R.

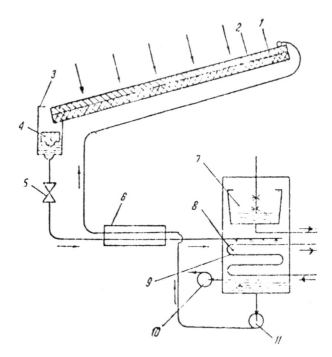

Source: AD-778 846, March 1974

In evaporator **7** water is cooled by partial evaporation to 7° to 12°C; passing through absorber **8** the water vapor is absorbed by strong lithium bromide or lithium chloride salt solutions. The yielded absorption heat is channeled by absorber coils 9 through the cooling water. The weak solution from the absorber is conveyed by pump 11 through heat exchanger 6 into the solar collector 1 for regeneration; the lower part of the solar collector has thermal insulation 2 which serves for regeneration of the solution.

Flowing in a thin layer, the solution is heated by solar radiation up to $45°$ to $60°C$, increasing its concentration by releasing water through evaporation. From there the solution flows back through a container 3 having a float 4 and a valve 5, into the absorber 8 to repeat the cycle. Any air entering the absorber through leakage or dissolved in fluid is eliminated by vacuum pump 10. The solar collector acting as regenerator can be in the form of an open basin with a heat-insulated bottom. To obtain cold water at $7°$ to $12°C$ temperature suitable for air conditioning the lithium chloride salts should be used, being less aggressive on metal and cheaper than lithium bromide salts (12).

The same institute in 1970 designed and tested a solar absorption refrigerating unit at Ashkhabad, Turkmen SSR, a variant of the system in Figure 3.29, using an open-plate generator and sprinkler chamber. In a dry and hot climate, this installation with sprinkler chamber is more advantageous than other models, as the heat losses in the cooling room are much larger than the capacity of the solar cooling installation. In addition, this system can work day and night (at night as an evaporating conditioner) and as such is recommended in regions with dry and hot climate with moderately cool nights.

This installation is composed of a solution regenerator, gutter, floating regulator, heat exchanger and sprinkler, absorber, evaporator, sprinkler chamber, vacuum pump, pumps, a check valve adjusted for uniform sprinkling of nozzles, drip pan, and a fan.

The solution regenerator, installed on roof of a building is faced toward the south with a $15°$ offset westward and has an area of 4.8 x 4.8 meter. The heat exchanger and sprinkler are made of double pipes; the heat exchanger has an area of 0.33 m^2, and the sprinkler (made of stainless steel) is 33 mm in diameter, with 3 mm diameter holes at 40 mm intervals. The absorber consists of 84 pipes (6 rows, 14 pipes each) with 21 x 2 mm diameter, and a total length of 760 mm. In general, the principles, operation cycle, and type of salt solutions are the same as those for the previous installation in Figure 3.16 (13).

References

(1) Löf, G.O.G., "The Heating and Cooling of Buildings with Solar Energy," *Introduction to the Utilization of Solar Energy*, McGraw-Hill, New York, 1963, 239-294.

(2) Thomason, H.E. and Thomason, H.J.L., Jr., "Solar House Heating and Cooling Progress Report," *Solar Energy*, no. 1, 1973, 23-39.

(3) "Meeting Reflects Solar Energy Optimism," *Chemical and Engineering News,* July 30, 1973, 14-16.

(4) Noguchi, T., "Recent Developments in Solar Energy Research and Applications in Japan," *Solar Energy,* no. 2, 1973, 179-187.

(5) Bahadori, M.N., "A Feasibility Study of Solar Heating in Iran," *Solar Energy,* no. 1, 1973, 3-26.

(6) Avezov, R.R., Zakhidov, R.A., Umarov, G.Ya. and Minchuk, V.I., "Results of Experimental Studies on the Joint Performance of a Heat-Pump Solar Unit of a Heating and Cooling Radiator System," *Geliotekhnika,* no. 5, 1970, 56-59.

(7) Oniga, T., "Absorption Cooling Unit with Fixed Conoidal Reflector," *Proceedings of the United Nations Conference on New Sources of Energy—Solar Energy: III,* Rome, August 21-31, 1961, vol. 6, New York, 1964, 41-50.

(8) Trombe, F. and Foex, M., "Economic Balance Sheet of Ice Manufacture with an Absorption Machine Utilizing the Sun as the Heat Source," *Proceedings of the United Nations Conference on New Sources of Energy—Solar Energy: III,* Rome August 21-31, 1961, vol. 6, New York, 1964, 56-59.

(9) Ashar, A.G. and Reti, A.R., "Engineering and Economic Study of the Use of Solar Energy Especially for Space Cooling in India and Pakistan," *Proceedings of the United Nations Conference on New Sources of Energy—Solar Energy: III,* Rome, August 21-31, 1961, vol. 6, 1964, 66-74.

(10) Yellot, J.I., "Solar Energy Progress—A World Picture," *Mechanical Engineering,* vol. 92, no. 7, 1970, 28-34.

(11) Farber, E.A., "Solar Energy, Its Conversion and Utilization," *Solar Energy,* no. 3, 1973, 243-252.

(12) Kakabayev, A. and Khandurdyev, A., "Absorption Solar Refrigerating Installation with Open Solution Regenerator," *Geliotekhnika,* no. 4, 1969, 28-32.

(13) Kakabayev, A. and Rakhmanov, A., "An Absorption Solar Cooling Installation with Sprinkler Chamber and Its Test Results," *Geliotekhnika,* no. 4, 1971, 38-43.

Simulated Systems

The following discussion is from PB-228 877, *Modeling of Solar Heating and Air Conditioning* by W.A. Beckman and J.A. Duffie of the University of Wisconsin, prepared for the National Science Foundation and issued July 1973. A particular residential building configuration has been selected for study and has been operated in the Albuquerque climate to meet heating, service hot water and cooling loads from a combination of solar energy and auxiliary energy, and represents one of many possible configurations of such systems.

System Models

The concept of system modeling has been to develop a driver program which will automatically connect the output of one component subroutine to the input of any other component subroutine. In this way, a complete system can be built from standard components (i.e., subroutines). The driver program gathers together all the necessary information and simulates the operation for any desired weather condition over any desired time period.

A system model has been operated as shown in Figure 4.1 (not shown are the control systems for the heating, cooling, and service hot water). The system is designed to meet heating, service hot water and cooling loads from a combination of solar energy and auxiliary energy, and represents one of many possible configurations of such systems. The major components of this system are as follows (numbers correspond to the numbers on Figure 4.1):

(1)   The collector is a flat-plate forced-circulation water heater using parallel flow upward through tubes which are thermally in good contact with the plate. The major design features of the collector

97

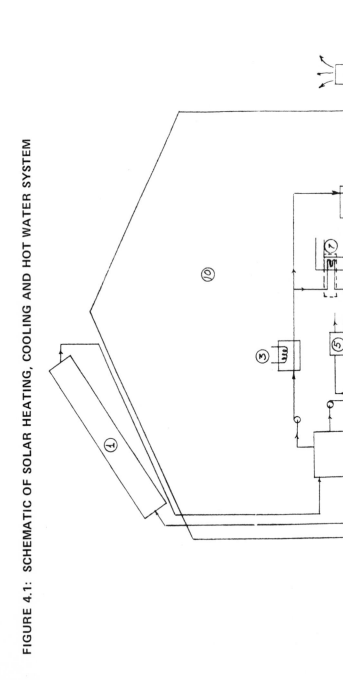

FIGURE 4.1: SCHEMATIC OF SOLAR HEATING, COOLING AND HOT WATER SYSTEM

Source: PB-228 877, July 1973

are included in three parameters which describe its operation: U_L, F_R, and the product $R\tau\alpha$. The overall loss coefficient, U_L, is dependent primarily on the number of covers, wind, speed and infrared emissivity of the plate. The overall plate efficiency factor, F_R, is dependent on collector geometry, fluid flow rate and U_L. The product $R\tau\alpha$ includes collector orientation, number and nature of covers, and solar absorptivity of the plate.

The collector is limited in temperature of its operation by maximum allowable system pressure. An upper limit of temperature can be set, with means provided to dissipate energy which would result in excessive water temperatures.

(2)   The storage unit is an insulated tank, either stratified or fully mixed. It is located inside the structure so that thermal losses add to cooling loads and subtract from heating loads.

(3)   A main auxiliary heater adds energy to the water leaving the top of the storage tank. It is a two stage heater that is controlled to off, low or high settings by the control system.

(4)   Service hot water is provided by a separate flow loop from the storage tank through a heat exchanger.

(5)   Service hot water is stored in a separate tank within the building.

(6)   A separate auxiliary energy supply is added directly to the storage tank to keep service hot water available at a temperature above a selected minimum.

(7)   Space heating needs are met by an exchanger which transfers heat from the hot water supply to the air in the building.

(8)   Cooling is provided by an absorption air conditioner, with the evaporator coil cooling and dehumidifying building air. The generator is heated by hot water from the storage tank.

(9)   A cooling tower is used to provide cooling water to the absorber and condenser of the air conditioner.

(10)   The building to be heated and cooled has a full range of energy and humidity gains and losses.

Figure 4.2 shows the nature of the control system assumed for heating and cooling, and the modes of operation of the system under various building temperature conditions.

↑ means as room temperature rises past the control temperature; ↓ means as room temperature drops past the control temperature. This corresponds to the 2 stage auxiliary energy source used in the Albuquerque simulation. In addition to these controls, there is another control which switches the system from heating cycle to cooling cycle or vice versa. The control of the single stage auxiliary supply for service hot water is separate but the same in function as the

FIGURE 4.2: SCHEMATIC OF CONTROLS AND MODES OF OPERATION

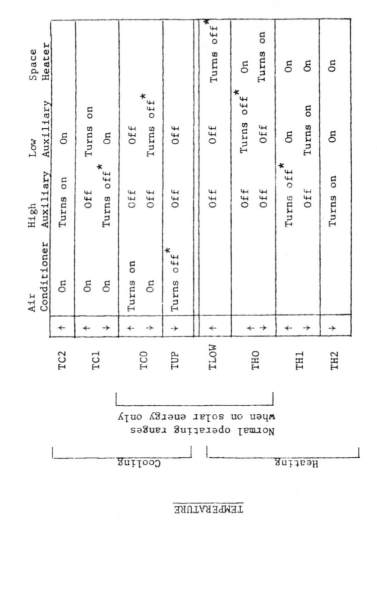

		Air Conditioner	High Auxiliary	Low Auxiliary	Space Heater
TC2	←	On	Turns on	On	
TC1	← →	On / On	Off / Turns off*	Turns on / On	
TCO	← →	Turns on / On	Off / Off	Off / Turns off*	
TUP	→	Turns off*	Off	Off	
TLOW	←		Off	Off	Turns off*
THO	← →		Off / Off	Turns off* / Off	On / Turns on
TH1	← →		Turns off* / Off	On / Turns on	On / On
TH2	→		Turns on	On	On

Normal operating ranges when on solar energy only

Cooling   Heating

TEMPERATURE

*Turns off if on.

Source: PB-228 877, July 1973

controls for heating and cooling. The model representing this system has been operated in the Albuquerque, New Mexico climate for the purposes of (a) proving the modeling and programming concepts, and (b) evaluating the thermal performance of this particular system. The main design variable considered to date has been collector area. The values of major design parameters assumed in this preliminary study are listed in Table 4.1.

TABLE 4.1: ALBUQUERQUE SIMULATION—MAJOR SYSTEM PARAMETERS

House Dimensions:	40 ft. x 45 ft., or 1800 ft.^{2} area.
Collector:	
Area	Variable, 150, 350, 650 and 950 ft.^{2} of 2-cover collector.
F_R	0.925
U_L	0.80 BTU/ft.^{2}-hr.-F
Orientation	Slope to south, 40° from horizontal.
Temperature Limit	230°F maximum.
Storage:	Variable but in fixed ratio to collector area, 12.5 lb$_m$/ft.^{2} of collector.
Air Conditioner:	Nominal 3 ton, operable within range of 0.35 to 1.15 of nominal capacity as temperature of hot water to generator and cool water to absorber and condenser vary.
Cooling Tower:	Water outlet temperature is taken as 10°F higher than ambient wet bulb, regardless of the load on the tower.

System Evaluation and Cost Analyses

The University of Wisconsin group is making preliminary studies to estimate the cost of delivered energy from the solar systems and to compare these costs with those associated with conventional systems. The approach to date is to make the thermal models complete and independent of the economic models, and then to evaluate the cost of delivered energy in terms of the estimated costs and life time of the systems.

After a system has been designed (i.e., numerical values selected for all design variables) it is operated for one year in a particular climate. During the year's operation all pertinent energy quantities are integrated. For the solar system the operating costs include fuel costs for the auxiliary heaters and electrical energy for operating additional pumps associated with the solar system. The yearly operating costs for a conventional system can also be calculated from the integrated energy quantities. The heating and cooling load for the particular building will be nearly the same for both solar and conventional operation (slight modifications are required to the integrated energy quantities due to the thermal losses from hot water storage tank in the solar system). The operating costs in the conventional system include fuel oil for the furnace and the service hot water and electrical energy for a standard vapor-compression air conditioner.

Since the solar system requires many of the same components as the conventional system, the cost comparisons between solar and conventional can be made on an incremented basis. That is, it can be assumed that the cost of a solar system, above the cost of a conventional system, includes an amount for pumps, pipes and controls, an amount for the absorption air conditioner over a vapor compression unit and an amount for the solar collector and storage tank.

In addition it is necessary to assume values of the interest rate, maintenance costs and the life time of the equipment. Since limited information is available on the life time of solar systems, it will be assumed that the conventional and solar systems have the same expected life-time and that both systems have the same maintenance costs. The values of the various parameters that are being used in the preliminary cost analysis for the simulated system are given in Table 4.2.

TABLE 4.2: PRELIMINARY COST ANALYSIS PARAMETERS

Cost of additional pumps, pipes, controls and modifications	\$ 250.00
Cost of absorption over vapor compression air conditioner	1000.00
Cost of collector on a per unit area basis	\$2 to \$4
Cost of storage tank per square foot of collector (\$.06/Lb$_m$ of H_2O stored with 12.5/Lb$_m$ H_2O stored/ft^2 of collector	\$ 0.75
Cost of money per year	0.08
Cost of electricity per KW-hr	\$0.02 to \$0.04
Cost of fuel oil per 10^6 BTU's	\$2 to \$4

Based on these costs, the costs of owning and operating the conventional system have been calculated, each above the base of the common costs. Examples of these comparisons are shown in Table 4.3. These numbers are all based on the Albuquerque climate.

It should be noted that the costs assigned to the solar collector are not based on detailed analysis of a particular collector design. Rather, the analysis says that if collectors cost C_C dollars per square foot, and if fuel cost is C_F dollars per 10^6 Btu, the comparisons of costs above the common base are as shown.

TABLE 4.3: COMPARATIVE COSTS

Fuel cost $/10^6$BTU	Collector cost $/ft^2$	Collector area corresponding to least cost solar system ft^2	Cost above base of the solar energy system $/yr	Cost above base of the conventional system $/yr
$ 2.00	$ 2.00	355	353	332
	4.00	140	396	332
2.50	2.00	480	378	377
	4.00	195	443	377
3.00	2.00	575	396	422
	4.00	285	482	422

Note: Fuel costs are dollars per million Btu's of delivered energy, taking into account furnace efficiency.

Comments

Selections of values of major design parameters for systems to be operated in various climates must be based on a combination of intuition, preliminary calculation, previous experience of others, or (where possible) good engineering practice. Many of the systems of interest have not been studied before, and it is not yet known how significant these parameters may be in determining costs of the systems. Optimization techniques are available for steady-state multiple variable systems, but because of large computing costs, new approaches may be required for the kind of transient operations implicit in solar energy applications.

MARSHALL SPACE FLIGHT CENTER STUDY

Background

The information which follows is derived from N74-26504, *The Development of a Solar-Powdered Residential Heating and Cooling System,* which covers work done by the George C. Marshall Space Flight Center in Huntsville, Alabama, prepared for the National Aeronautics and Space Administration and issued in May 1974.

The current worldwide shortage of petroleum dramatically emphasizes the need for alternative energy sources. Among the possible alternative energy sources, the most pollution-free, nondepletable, boundless source of all is solar energy. It is expected that the sun will ultimately be harnessed to produce electricity, synthetic liquid and gaseous fuels, and high temperature thermal energy for industrial processes. However, the first major application of solar energy will be to heat and cool buildings since this application will require the fewest technological advances and the least expenditures of time and money. Efforts are under way to demonstrate the engineering feasibility of using solar energy for this purpose. These efforts have been concentrated on the analysis, design and test of a full-scale demonstration system which is under construction at the National Aeronautics and Space Administration Marshall Space Flight Center, Huntsville, Alabama.

The basic solar heating and cooling system under development utilizes a flat plate solar energy collector, a large water tank for thermal energy storage, heat exchangers for space heating and water heating, and an absorption cycle air conditioner for space cooling.

Using previously developed computer tools, a wide range of solar energy collector studies has been conducted. The effects of collector design and operating conditions upon collector performance have been explored extensively. Thermal analyses of the energy storage system have been conducted. Numerous parametric studies of the total system performance have been conducted using a sophisticated system simulation computer program. This program uses measured meteorological data and mathematical models of all system components to perform a transient energy transfer analysis through an entire year.

Practical methods of effecting automatic control of the overall solar heating and cooling system have been investigated, and pressure distribution studies also have been completed. Incorporated into the test system described below are the results of these studies.

Description of Test System

The test system is made up of several major subsystems as shown in Figure 4.3. These are the solar collector, hot water fired absorption air conditioner, heat storage tank, air conditioning system cooling tower, and three house trailers to simulate living quarters. The working fluid is deionized water pressurized to 30 psig to prevent boiling. The system operation is controlled automatically. This allows continuous unattended operation.

The heart of the test article is the solar collector. It consists of 31 segments; each segment is made up of an aluminum tray containing 7 heat exchangers. The heat exchangers are 1100 series aluminum coated with an electroplated frequency selective coating that allows relatively high efficiency collection of solar energy at temperatures up to 260°F (Figure 4.4). The heat exchangers are effectively insulated on the front face by a Tedlar wire composite. The Tedlar allows transmission of the solar energy, but greatly reduces convective heat losses from the heat exchanger.

The Tedlar cover also functions as a greenhouse, trapping solar energy. The wire grid serves as structural support for the Tedlar. The back of the heat exchanger is insulated by 6 inches of fiber glass household insulation. The heat exchangers

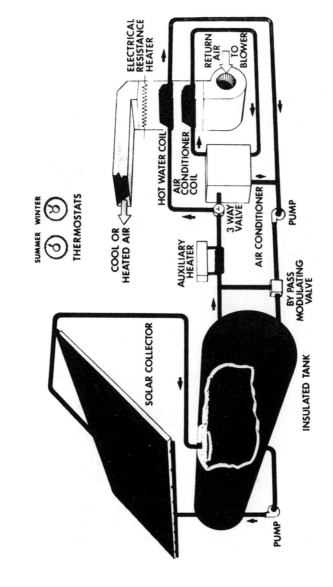

FIGURE 4.3: SYSTEM SCHEMATIC

Source: N74-26504, May 1974

FIGURE 4.4: MSFC COLLECTOR DESIGN FEATURES

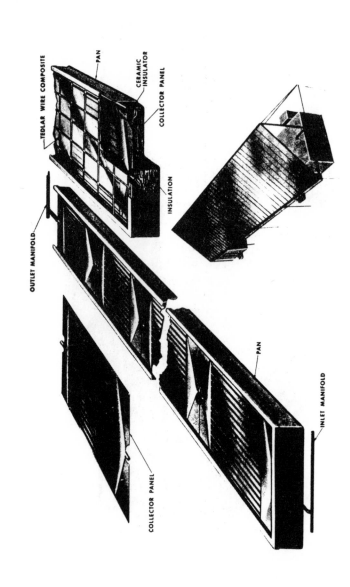

Source: N74-26504, May 1974

are held in the aluminum trays by wooden strips which form channels on two sides of the exchangers. Teflon clips prevent the edges of the heat exchangers from touching the sides of the aluminum trays, and thus prevent thermal shorts. The fluid entrances and exits of the 31 segments are connected to manifolds that transport deionized water to and from the thermal storage tank. The collector system weight is 4,600 lb, providing a roof loading equivalent to approximately 20%.

The trailer complex, which makes up the dwelling to be cooled and heated, is constructed of three surplus, office-type house trailers, two 10 ft by 58 ft and one 12 ft by 30 ft, secured together, with interconnecting doors and floor located air distribution ducting. The complex is positioned with its long dimension parallel to an east-west line in order that one face of the roof is facing due south so that the maximum solar energy input is possible. The nature of house trailer construction does not provide sufficient strength to support a roof structure of even minimum structural requirements; therefore, a free standing roof was constructed over the trailers to support the solar collector trays and associated plumbing. The roof is supported on fourteen 6-inch-diameter steel posts set in concrete, has a slope of 45°, and utilizes trussed rafter construction with rafters placed on 24-inch centers.

The roof slope of 45° has been found to be the slope condition that provides a maximum amount of solar input for all seasons of the year. The trailers have an area of 1,500 ft^2 and the expected heat loss is equivalent to a conventionally constructed house of approximately 2,500 ft^2.

The thermal storage subsystem is shown in Figure 4.5. The main components are a 4,700 gallon aluminum storage tank to contain 3,600 gallons of deionized water, pump valves and controls, a wooden house for weather protection, and household fiber glass insulation blown between the tank and interior of the house. The tank is from surplus and is significantly oversized from a cost optimization standpoint. The storage subsystem and the solar collector form the two major components of the heat collection and storage loop. The other fluid loop connects the absorption air conditioner and the thermal storage subsystem.

Cooling is provided to the trailer complex by a lithium bromide/water absorption air conditioner. The specific unit was commercially manufactured and sold through 1967 by Arkla. The unit provides approximately 3 tons of cooling at a coefficient of performance of 0.67. The unit has been modified by removing the gas-fired burner and replacing it with a water heat exchanger. Hot (210°F) water from the thermal storage system is pumped through this heat exchanger to drive the absorption cycle air conditioner. An auxiliary electrical heater is installed in the water loop to insure hot water in the event the thermal storage system becomes inadequate.

The lithium bromide/water cycle operates under a partial vacuum, approximately 100 mm Hg pressure in the generator and 8 mm Hg pressure in the evaporator. The refrigeration effect is obtained by water evaporating at the low pressure which provides an evaporator temperature of approximately 45°F.

FIGURE 4.5: THERMAL STORAGE TANK

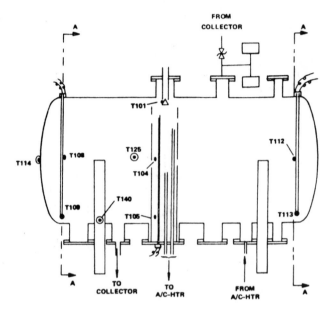

●= Internal thermocouples   ⊙ = External thermocouples
T1 series = Thermocouples symmetrically located on opposite side or end

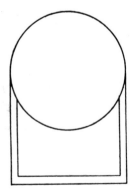

View **A-A**
External thermocouples, north end

Source: N74-26504, May 1974

Heat rejection from the air conditioner is accomplished by a forced draft cooling tower. The cooling tower rejects heat by evaporating water into an air stream created by an electrically driven blower. The cooled water is circulated through the air conditioner adsorber and condenser to remove the heat of absorption and heat rejection by the fluid cycle. Heat required within the trailer during winter months is supplied by bypassing hot water from the storage tank around the air conditioner to an air/water heat exchanger located in the air distribution duct. If the water temperature in the storage tank becomes inadequate, supplementary heat can be added by electrical heaters in the air distribution duct.

Modeling Techniques

To evaluate the effect of different design options upon the performance of the total solar-powered system a computer program has been developed and refined over a 2.5 year period to simulate the transient behavior of the system for an entire year. A brief description of modeling techniques and results of simulations is presented in the following paragraphs.

Figure 4.6 shows the basic components which make up the simulation program. The six subsystems which comprise the total system are each separately modeled. Environmental data in the form of time-varying ambient temperature, wind speed, and direct and diffuse solar fluxes are required inputs. These environmental data are required for an entire year to allow transient analysis over the entire year. When time-varying solar data are not available, as is usually the case, techniques for calculating these data from available, whole-day, total insolation values have been developed.

The method which is used to recorrelate the whole-day flux totals into time-varying fluxes is to simply determine the average atmospheric attenuation of the solar constant for each day and apply this attenuation factor to the solar constant throughout the day while conducting the system analysis. Thus the solar flux intensity is held constant through the day, while the relative position of the sun to the collector (the incidence angle) is calculated instantaneously throughout the day, based upon input values of latitude and collector tilt angle. A control system logic routine allows for choices in such parameters as thermostat settings, energy exchanges between components, flow rates, switches, etc.

The six subsystem models used in the computer program to simulate the MSFC solar-powered facility are described below. These models pertain only to the residential solar-powered system being developed at MSFC. Other systems for other buildings in other regions would require different models, although the same simulation approach should prove valuable for analyzing these other systems.

The six subsystem models are very simple, with the exception of the solar collector model, which is a sophisticated mathematical model that allows accurate determination of the transient thermal performance of the collector.

FIGURE 4.6: BASIC COMPONENTS OF THE SIMULATION PROGRAM

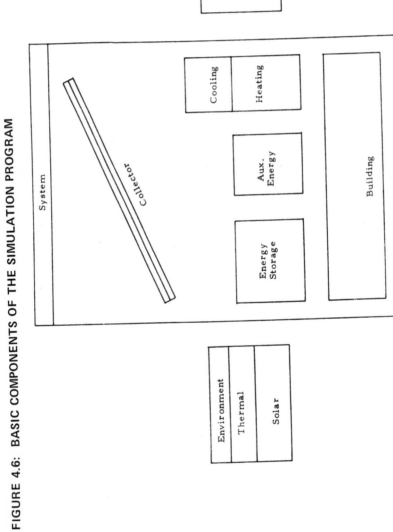

Source: N74-26504, May 1974

Building Model: The building is treated as a fixed thermal capacitance connected to the outside environment through a variable thermal resistance. The capacitance is assumed to be 10,000 Btu/°F. This capacitance value represents the effective thermal inertia (mass times specific heat) of the building and its contents, which serve to dampen the effect of external temperature variations on the inside air temperature. The resistance value is different for hot weather than for cold weather.

When the ambient temperature is above 70°F, the resistance value is set at 0.0007°F hr/Btu, which corresponds to a 3-ton (36,000 Btu/hr) cooling load at 95°F ambient temperature. When the ambient temperature is below 70°F, the resistance value is set at 0.00117°F hr/Btu, which corresponds to a 60,000 Btu/hr heating load at 0°F ambient temperature. The reduced resistance value for cooling is used to include latent loads which are important during the cooling season. These resistance values were based upon MSFC load calculations, and represent the overall heat transfer path between the inside air and the outside air.

Energy Storage Model: The energy storage system is treated as a fixed thermal capacitance for the MSFC simulations. The capacitance value is an input quantity which represents the mass times specific heat product of the water in the energy storage tank. The storage system heat losses are treated as a fixed thermal resistance, and heat flows through this resistance to the outside environment. A maximum temperature is also a required input.

Auxiliary Energy Model: The auxiliary energy system is a zero capacitance thermal energy source. Several options are available concerning the control of the auxiliary energy system, including whether the heat is added to storage or used directly to power the heating, cooling, and water heating components. The output heating rate of the auxiliary heater is 60,000 Btu/hr for heating and 55,000 Btu/hr for cooling.

Heating Model: The space heating model is a zero capacitance input/output unit which utilizes heat from either the collector, energy storage, or the auxiliary energy system, depending on control logic selection and instantaneous conditions. The efficiency of the unit is an input variable, being defined as heat output divided by heat input. The heating rate output as a function of water temperature and air temperature as these fluids enter the best exchanger is also input to simulate the actual heat exchanger performance.

Cooling Model: The space cooling model is a zero capacitance heat-driven air conditioner, with its performance defined by an input coefficient of performance (0.65). This input/output unit may be driven with heat from either the collector, storage or auxiliary, depending on prevailing conditions and control logic selection. The cooling rate is considered constant at 36,000 Btu/hr.

Solar Collector Model: The solar collector model is a flexible energy transfer model that allows considerable variations in collector design. Input design variables include collector area, tilt angle, number of transparent covers, characteristic dimensions and spacings, and thermophysical properties of absorber plate, transparent covers, and backside insulation. Water is assumed as the energy transport fluid and its flow rate is an input parameter. The thermal capacitances of the plate, covers, insulation, and water are all used in the transient numerical treatment of the collector. The accuracy of the collector model is well verified by test data.

Simulation Results

The computer program takes the inputs and models described above and conducts a transient energy transfer analysis through the year, obtaining all pertinent energy flows including energy used for heating and cooling, energy collected, auxiliary energy used, and energy wasted for lack of storage capacity. Daily totals and cumulative totals throughout the year are determined for all of these energy flows.

This simulation computer program has been widely used to compare different system designs. The major parameter of comparison has been taken as the dimensionless ratio, Q_{aux}/Q_{tot}, which represents the fraction of the total energy requirements for heating and cooling which must be met with conventional, auxiliary energy. Obviously, the smaller this value is, the better the system performance is. One comparison is presented in Figure 4.7. This comparison is between two collector designs and two tank sizes, all as a function of collector area. The benefits of using two Tedlar covers and large storage are apparent. Details on the effects of varying other system parameters for the simulated studies are included in the report.

FIGURE 4.7: EFFECT OF COVER DESIGN ON SYSTEM PERFORMANCE

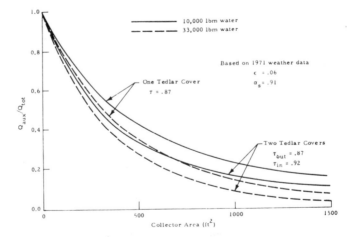

Collector Area (ft^2)

Source: N74-26504, May 1974

Conclusions

The following conclusions are drawn, based upon the results of studies and tests. The utilization of solar energy for space, heating and air conditioning is technically feasible with properly applied current technology. The best solar heating and cooling system design should include:

A solar energy collector with an integral tube-in-sheet absorber plate coated with a selective surface having high solar absorptance and low infrared emittance.

A sensible heat storage system with water as the storage medium (the same water should be used as the energy transport fluid to and from the collector and output units).

An absorption cycle cooling system.

An auxiliary heat source that utilizes storable fuel (fuel oil, propane, etc.) rather than natural gas or electricity. This will prevent peak, simultaneous drains on the conventional utilities by solar-powered installations whenever several days of cloudy weather occur.

The use of computerized simulations of solar system performance is essential to compare different system designs, to properly size collector and storage, and to generate near-optimum designs. With increases in solar energy usage and technology, system design from handbooks should be possible.

UNIVERSITY OF PENNSYLVANIA STUDY

NSF-RA-N-73-005, *Conservation and Better Utilization of Electric Power by Means of Thermal Energy Storage and Solar Heating, Final Summary Report,* by M. Altman, H. Yeh and H. Lorsch, of the University of Pennsylvania, prepared for the National Science Foundation and issued July 1973 describes a project in which the technical and economic aspects of solar heating, off-peak air conditioning, and thermal energy storage were investigated. A solar heating system, which incorporates a conventional heating system as a back-up, was designed.

Performance calculations were made on a simplified model of this system built into a townhouse in Washington, D.C. Economical collector and storage sizes were determined. Three facilities for the testing of solar collectors were designed built, and successfully operated. Twenty-four collector configurations were considered, and samples of the better ones were tested in the laboratory and under actual outdoor insolation conditions. Two configurations were identified which promise significant improvements in cost effectiveness over existing collectors.

Two off-peak air conditioning systems were designed, and system prototypes were built and tested. One of these, the refrigerant recondenser system, would be economically attractive for the operator of a medium-sized office building, but it would not be for a homeowner. Theoretical and experimental work was performed on latent heat energy storage materials. Criteria for suitable materials were developed; the reasons for the disappointing performance of previously used materials were established, and promising new materials were identified. Exploratory tests confirmed their favorable properties. Economic and marketing considerations for solar and off-peak space conditioning were explored. The effects of the introduction of these new systems on different economic sectors was tentatively evaluated. Market research was performed, and suitable marketing strategies were defined.

Feasibility Studies
for Large Scale Applications

NATIONAL SCIENCE FOUNDATION STUDY

The material in this section is from a report, *Subpanel IX—Solar and Other Energy Sources,* by A.J. Eggers, Jr. of the National Science Foundation, prepared for the Atomic Energy Commission and issued in October 1973.

Overview

The sun is an inexhaustible source of an enormous amount of clean energy available nearly everywhere in the world. The technical feasibility of using solar energy for terrestrial applications is well established. On the other hand, solar energy is diffuse (17 w/ft^2, 24 hour average in the U.S.) and variable (from zero to a maximum and back to zero each 24 hours). These two factors of low energy density and variability, combined with the ready availability of inexpensive fossil fuels, have until now, discouraged the development of systems suitable for widespread use.

However, a study conducted by leading university, industry and government experts (*Solar Energy as a National Resource,* NSF/NASA Solar Energy Panel, December 1972) concluded that a substantial development program could achieve the technical and economic objectives necessary for practical systems. In certain areas, practical systems are already in operation, e.g., domestic hot water heaters, remotely located buoy power systems, house heating systems, and waste conversion plants.

Significance and Benefits: At an average energy conversion efficiency of 5%, less than 4% of the U.S. continental land mass could supply 100% of the Nation's current energy needs. Thus, solar energy could contribute significantly to the national goal of permanent energy self-sufficiency while minimizing environmental degradation.

Although the full impact of solar energy probably won't occur until the turn of the century, the economic viability of several of the applications, e.g., heating and cooling of buildings (HCB), wind electric power (WEP) and bioconversion to fuels (BCF) could be developed and demonstrated in the next 5 years. Ultimately, practical solar energy systems could easily contribute 15 to 30% of the nation's energy requirements.

In most cases the development of practical systems will not require high technology. Thus, the research and development costs for solar energy should be very small in relation to the value of the energy saved. Current estimates indicate that the value of the fossil fuel to be saved in one subprogram area alone, i.e., HCB, would equal the cost of an entire accelerated ($1 billion) R&D program 7 years after practical systems become commercially available.

Program Plan Summary: Approximately 25% of the nation's current energy consumption is used for HCB purposes. For maximum utilization, solar HCB systems including domestic hot water systems suitable for both new construction and existing buildings must be investigated. Ultimately, 30 to 50% of the national heating and cooling energy requirement could be furnished by solar energy with the accompanying benefits of fuel savings, reduced pollution, and independence from complex energy transmission and distribution systems.

No major technical barriers exist to prevent the development of practical systems. Most of the technical problems to be solved involve the development of low-cost per unit of energy components. Specifically, the collectors and cooling systems require the most improvement, but appear amenable to a determined development program.

There are major uncertainties with regard to public acceptance, legal rights to unshaded sun, the establishment of a supporting industry, and methods of marketing and financing high, first cost solar HCB systems. However, it seems probable that with the development of economically competitive systems, the benefits associated with solar HCB systems will provide major incentives for solving or accommodating these problems.

Introduction

Heating, cooling, and domestic hot water needs of institutional, industrial, and residential buildings can be met by using solar energy with existing technology. Achieving commercial availability for such systems by 1979, coupled with user acceptance, for large parts of the United States requires a development effort and demonstration of system performance. Proof of concept experiments will demonstrate systems performance and acceptability in various geographic locations. Several types of heating and cooling systems are envisioned.

The objective of the heating and cooling of buildings program is the achievement of commercial availability and widespread use of solar energy systems. Problems which must be addressed are developmental, economic, and societal. Development of components (energy collectors, refrigeration subsystems, storage subsystems, and conversion machinery) will result in lower costs, improved performance, and increased societal and commercial acceptability. Proof-of-concept experiments using government and privately-owned residential structures will result in demonstrations of improved economics and will serve as open laboratories to persuade industry and the public of the profitability and utility of solar energy systems. These experiments will be carefully designed to provide the data required by architects, system and component manufacturers, financers, builders, owners, and operators.

The program is structured to provide: (1) component research to increase the number of solar system combinations which will permit the greatest potential for meeting the varied requirements of building types in differing climatic regions of the United States; (2) point demonstrations of custom built systems for design, construction and operational experience and performance data; (3) proof-of-concept experiments (pilot plants) on a selected set of building type/system type/region combinations throughout the United States to establish the viable range of systems applications and compile performance and life cycle cost data; and (4) small solar communities, perhaps initially on Federal installations, to demonstrate the potential economics of scale and increase operational experience.

Demonstration in privately developed small communities would then be undertaken with appropriate underwriting of solar equipment incremental costs.

Status of Technology

Space Heating: A number of experimental solar-heated structures have been built and operated. Some were laboratory models and others were residences. Economic studies show that in a wide variety of U.S. climates, solar heating could now be competitive with electric heating. Architects are becoming interested in commercial construction of solar-heated homes and commercial buildings. However, subsystems, and building solar systems integration data required by architects and HVAC designers are not available.

Each building is a development project. Also, there is not sufficient component and building solar systems design, integration, and operational experience to convince the building industry and the public that solar heating is economic and reliable. The lack of a demonstrated solar cooling system also hinders implementation of solar heating because many potential users would like a heating system to which a cooling system can be readily added.

Space Cooling and Humidity Control: Solar cooling and humidity control is in the early development stage. Development of solar cooling and humidity control systems, particularly for hot humid areas, will be more expensive than the development of solar heating systems. The air conditioning equipment companies are reluctant to make the required investment before a market is assured and government funding is required. There are two classes of cooling and humidity control problems; one pertains to hot dry climates and the other to hot humid climates. Systems for hot dry climates are easier to develop and earlier implementation is possible. Combined solar heating and cooling systems for hot dry areas look economically promising now but component and system development and demonstration data are not sufficient to permit use by the building industry and the public. Climate data showing where such systems are feasible are also not available or not compiled in a useful way. The adaptation of air conditioning systems (e.g., evaporative, dessicant, night radiation, absorption, compressor, and heat pumps) to solar energy use is technically feasible but with more effort required for the systems for hot humid areas.

Implementation

No research barriers are delaying the implementation of solar heating and cooling of buildings. Rather, development is required to make systems economically attractive. Systems studies, using widely different climatic conditions and building requirements will specify the range of performance needed from the various components. An array of designs, making most efficient use of collected energy, will be developed for the various applications.

Thus, component parts development, namely collectors, energy storage and refrigeration systems using low grade thermal energy, and heat exchangers suitable for solar application will remove the barriers to applications. Problems of obtaining widespread acceptance are complex. The high first cost of solar equipment is incompatible with current marketing practices which emphasize low acquisition cost. Demonstration projects, with fully reported details are required to prove system economies, aesthetics, performance and reliability.

Rights to sun angles must be guaranteed by legislation or an investor cannot risk the possibility of construction adjacent to his building site obstructing the sun. Detailed design studies must be made to prove that new solar systems designs meet safety and health requirements. Planning commissions, coding authorities, and local trade unions must be convinced and modify appropriate regulations and practices which may interfere with solar powered building construction.

A national solar heating and cooling program is being coordinated by the NSF. Several other government agencies including AEC, GSA, NASA, HUD, NBS and DOD are working on various parts of the program. Various universities, companies, and individuals are conducting studies in all phases of the use of solar energy for heating and cooling of buildings. The objective is the widespread utilization and availability of solar energy systems for all building types and geographic regions in the United States. Proof-of-concept experiments, initiated in fiscal year 1974 will be carried into subsequent phases of the effort with emphasis on system definition and critical subsystems research.

Benefits of Implementation

The benefits of implementing a program to develop solar heating and cooling systems are many and include reduction of conventional energy consumption with attendant reduction of environmental pollution, decreased dependence on foreign fuel sources and a potential export market. The supply, marketing, and consumer sectors involved with solar heating and cooling are the same as those sectors concerned with conventional space heating and cooling and hot water functions. The estimated energy used for these functions in 1985 is projected to be 17×10^{15} Btu/yr.

This represents a limit to the energy that can be replaced by the application of solar energy to completely satisfy these functions. Forecasting an ultimate goal of 50% saving of fuels and electricity used through economically feasible applications of solar energy to space heating and cooling of buildings and hot water service, then the equivalent of 8.5×10^{15} Btu/yr can be saved. The application of solar energy to these functions is not a simple substitution of one fuel for another or for electricity, but requires important changes in construction technology and institutional concepts.

Certainly the implementation of this program will, in the long term, result in a reduction of fuel imports from abroad. With 25% of total national energy consumed now going into heating and cooling of buildings, achieving a 10% application of this technology would result in a 2.5% reduction in fuel consumption. In particular, however, it could have its greatest impact in high fuel cost regions.

Reliability of solar energy utilization systems should be high in the sense that the supply of energy cannot be limited by political or economic processes, either local or foreign. Lack of sunlight for periods of time can be mitigated through the use of storage systems.

From an environmental standpoint, solar energy for heating and cooling adds no pollutants or thermal emissions. Solar energy does not disturb the balance of energy on earth and is environmentally nondegrading.

In summary, the benefits of implementing solar energy utilization for heating and cooling of buildings include increased efficiency of energy utilization and increased independence from fossil fuel sources.

Economics of Implementation

The estimated price of delivered solar energy for space heating and cooling and water heating depends on its location. Ultimately, viable solar systems will offset their purchase price by fuel savings in a reasonable time period.

Optimal systems will provide between 40 to 60% of the heating and cooling requirement for those homes (and buildings) installing the systems. The remaining requirements will be satisfied by conventional energy sources.

Implementation of the system will have a favorable effect on the labor market. The technology and production knowledge exist for all components of the system. Collector technology will permit use of mass production techniques. Installation of systems will not require special skills. Some minimal amount of training will be required.

The energy crisis is worldwide; therefore, development of an economically viable system will result in potential markets for the export of technology or capital goods to other countries after a U.S. market is developed.

Impact of Implementation

There do not appear to be any substantial negative impacts resulting from the utilization of solar energy. The materials of construction exist and are commonly used by the industry. These include glass, thermal insulating materials, plumbing materials, working fluids, and thermostatic control systems. Application of these materials to solar energy utilization will not compete with other uses since this will supplement existing heating systems. Legal or regulatory restrictions do not appear to exist.

Work proposed under this program should have a slight but favorable effect on employment. However, if solar heating and cooling of buildings were to achieve wide-scale acceptance and application, substantial changes in the equipment and building industries might occur. An export market may also develop although it can be expected to lag behind the domestic market.

There are specific benefits which result from the application of solar energy to heating and cooling. Existing basic technology is used with minimal impact on existing installed systems, permitting both new installation and retrofit. Next, there is a decrease in pollution which results from the installation of these systems although there is an increase in net useful work. Finally, the use of solar energy offers independence for at least some of man's needs from depletable energy sources. Although these goals can be realized although to only a small extent by 1980, the nucleus of the business would be established (many years ahead of the expected time required for the normal market development).

MITRE STUDY

Background

Currently a number of individual houses in the U.S. are provided with solar heating systems. Experimental structures have been built and operated to various extents, but comprehensive technical and economic data have not been collected and evaluated on most of these structures. Once systems employing efficient solar energy collectors, converters, and storage devices are developed, at costs competitive with conventional systems, social and institutional problems may still inhibit widespread construction of solar heated and cooled buildings.

No technological breakthroughs are needed to construct and operate systems of these types. What is needed most, at this point, is research on engineering of components and total building systems to make them more cost-effective, research on manufacturing processes that can achieve cost savings from mass production, and the development of innovative marketing concepts. In addition, concern must be generated in the building industry for making maximum use of solar energy in the heating and cooling of buildings.

In a systems analysis carried out by MITRE Corporation, PB-231 142, *Systems Analysis of Solar Energy Programs,* completed in December 1973 for the National Science Foundation preliminary systems evaluation has been conducted to indicate the potential future applicability, costs and benefits of solar energy systems that are under consideration for research support by the National Science Foundation.

Systems Evaluation

The results of this evaluation as applied to systems for solar heating and cooling of buildings are given in Table 5.1. All costs and prices are in terms of 1970 dollars.

TABLE 5.1: SOLAR HEATING AND COOLING SYSTEMS EVALUATION

CRITERION	ASSESSMENT
1. Possible Impact on Energy Requirements	
a. Types of Products	Space-heat; air-conditioning; hot water.
b. Types of Users	Residential, commercial, industrial, farm and other buildings.
c. Siting Restrictions	No restrictions. Systems will be integral parts of buildings, throughout U.S.
d. Potential Capabilities	Could supply about 40% of requirements for heating and cooling of buildings, or about 5% of U.S. energy requirements, i.e. about 1.0×10^{16} BTU/year, by the year 2020; economically, competitive system could be intially available within 5 years; could be produced at sufficient rate to supply systems for new housing and for some retrofits; constraints are mainly those of public and institutional acceptability.

(continued)

FIGURE 5.1: (continued)

CRITERION	ASSESSMENT
2. Technical Characteristics	
a. Flexibility	Mismatches in load and supply will require supplemental energy supply, particularly in Northern Regions of U. S.
b. Technical Reliability	Should be highly reliable.
c. Sensitivity to Weather	Affected by clouds, as well as by fog, haze, snow, ice, etc., but no high risk of disruption if energy storage or supplemental energy supply are included; same sensitivity to natural disasters as buildings in which systems are installed.
d. Security	May have some vulnerability to vandalism or malicious mischief.
e. Base-Load/Peak-Load Capabilities	Should be able to meet base-load and peak-load demands with energy storage or supplemental energy supply.
f. Storage Requirements	Supplemental energy supply systems will almost always be used, if available, since it is more economical than extended period storage.
g. Interfacing Requirements and Capabilities	Will require interfacing with supplemental energy supply; no interfacing capability requirements problems foreseen.
h. Redundancy	Supplemental energy supply required.
i. Synergistic Effects	Could be used in combination with any of the other solar energy systems to reduce pro rata costs or increase performance and operational reliability of the systems.
j. Transmission and Distribution	None, except within building, via pipes and ducts
3. Potential Economic Viability of System	
a. Capital Costs	Capital cost, by 1985, of combined solar-plus-supplemental-system for heating and cooling of typical residence estimated at about $8000 versus $3500 for conventional system. Lifetime comparable to conventional system.
b. Operating Costs	Operating costs, by 1985, estimated to be about $100/year for solar-plus-supplemental system (above) versus about $400/year for conventional system. Plant factor of collector system about 10% to 25% depending on Region.

(continued)

TABLE 5.1: (continued)

CRITERION	ASSESSMENT
c. Economies-of-Scale	Control system costs will be a small fraction of total system costs, so economies-of-scale small.
d. Estimated Price of System Products	Total annual heating and cooling costs (including amortization of equipment) estimated at about $1000/year, by 1985, for solar-plus-supplemental system, versus about $800/year for conventional system.

4. Environmental Impacts of System

a. Environmental Effects	Insignificant
b. Land-Use Requirements	Insignificant, since systems will normally be integral parts of buildings they serve or in the case of large buildings, surrounding land (e.g. parking lots).
c. Water Requirements	Very little water required for space heating. That used would be recycled. Normal amounts of water used for hot water supply.
d. Heat Balance	No significant change in the heat balance of the environment is anticipated, from the use of this type of system.
e. Visual, Noise, Odor Effects	Roof tops oriented southward and absorbent structures may cause visual problems. Some types of A/C units may produce noise or odor.
f. Ecological Problems	None.
g. Plant Failure	May be occasional problems associated with working fluid leaks.

5. Sociological Acceptability of Systems

a. Esthetics	Architectural appeal a very important consideration. Requirements for tree removal may be a problem at many possible installation sites.
b. Noise, Odor, etc.	See 4e above.
c. Safety	No serious problems foreseen.
d. Area Impact	No problems foreseen.

(continued)

TABLE 5.1: (continued)

CRITERION	ASSESSMENT
6. Institutional Constraints	
a. Institutional Acceptability	Conservative nature of building construction, financing, regulations, and operating industry makes innovation difficult. Major effort required to develop acceptability.
b. Implementability	Mass-production of system components needed to reduce capital costs. Incentives of various types, e.g. lower interest rates, faster amortization rates, lower property taxes, etc. for householders and other building owners who utilize solar systems for the heating and cooling of their buildings.
c. Zoning	May require three-dimensional (3-D) zoning restrictions to prevent new buildings or other structures from shading existing buildings with solar heating or cooling systems.
d. Ownership	Probably the building owners would own these types of heating and cooling systems. Household systems may require maintenance services supplied by utility companies or others.
e. Legal Problems	None foreseen, except those arising from 3-D zoning problems such as shading of buildings, trees, etc., and removal of trees and other vegetation.

Source: PB-231 142, December 1973

General Assessments

Systems for heating and cooling of buildings have proved to be technically feasible, although their capital costs are still considerably higher than conventional heating and cooling systems.

The total annual cost of solar systems, including auxiliary conventional system fuel costs, O&M and amortization of equipment, is still about twice that of conventional heating and cooling systems. This ratio should decrease as system costs decrease and fuel prices rise (i.e., by 1985, the annual cost of solar heating and cooling systems may only be about 30% higher than conventional systems).

Mass production of solar heating and cooling units should be undertaken, as soon as possible, in order to reduce capital costs.

In addition, various types of financial and tax incentives should be provided. These would include, among other incentives:

lower interest rates

tax breaks

This would serve to encourage householders, and others, to save fossil fuels by using solar systems for heating and cooling of buildings.

Systems for solar heating and cooling of buildings need proof-of-concept experiments that will demonstrate their applications and encourage their utilization.

Problem Areas

The solar heating and cooling of buildings has been categorized in terms of problem areas. In categorizing, the problems have been sorted in terms of whether they relate to the objectives of demonstrating

(1)   the technical feasibility

(2)   the economic viability

(3)   the environmental desirability

(4)   the sociological acceptability, or

(5)   the institution constraints of the system.

Then the problems, so identified, have been broken down, further, to specify whether they pertain to

(1)   the collection subsystems

(2)   the conversion subsystems

(3)   the storage subsystems

(4)   the transportation and/or transmission and distribution subsystems, or

(5)   the utilization subsystems of the system of concern, or whether they are problems pertaining to

(6)   systems integration.

Table 5.2, on the following two pages, presents the problems for the heating and cooling of buildings.

TABLE 5.2: PROBLEM AREAS

System Component	Technical	Economic	Environmental	Sociological	Institutional
Collection	Limits on Q/E Efficiency vs temperature Space required Sensitivity to weather, dirt, vandalism Leaks Combined thermal & electric Reliability, life Mass production techniques Tracking systems	2/1 reduction in cost ($2/ft^2) Maintenance Life-cycle costs	Glare Leakage of working fluids Impacts of solar buildings on trees, etc.	Aesthetic appearance Safety Cleaning and maintenance Vandalism	3-D zoning Roof access limitations Building codes
Conversion	Absorption air conditioning Compression air conditioning —engine type Efficiency of heat exchange Night-sky radiation Dehumidification	Initial cost Maintenance Standby system Life-cycle costs	Heat rejection —disposal of condensate	Public acceptability of A/C systems	Underwriter codes
Storage	Medium selection Architectural compatibility Control complexity Material degradation Contamination/cleaning Leaks, sealants Operating temperatures	Initial cost Maintenance Packaging Cost/availability of media Life-cycle costs	Toxic substances Water use	Safety, confinement Psychological (fear)	Building codes

(continued)

TABLE 5.2: (continued)

System Component	Technical	Economic	Environmental	Sociological	Institutional
Transportation and distribution	Control mechanisms	Control mechanisms	No problems foreseen	No problems foreseen	No problems foreseen
Utilization Systems integration	Optimize appliances Design of integrated systems Standby requirements System optimization O&M design Test and evaluation procedures Marketing analysis Technical plans Dissemination of technical information	Life cycle cost of appliances Life-cycle costs of integrated systems Economic analysis of system tests Determine economic trade-offs to optimize system parameters Assess economic impact on industry, etc. Economic analysis for marketing studies Means of financing Dissemination of economics information Funding requirements for technical centers	Impact of appliances Develop standardized procedure for measuring environmental impacts Determine environmental impact of systems Assess environmental acceptability Environmental impact codes Dissemination of environmental impact data	Impact of appliances Determine public acceptability Coordinated plans for measuring public acceptability Educate public on solar heating and cooling	Regulations for use of appliances Institutional acceptability Technical standards Building codes 3 dimensional zoning Architectural and engineering education Information dissemination Union and trade acceptance Government and financial organization acceptance Methods of removing constraints

Source: PB-231 142, December 1973

No attempt has been made to weigh these problem areas in terms of importance. It will be noted that a majority of the problems that have been identified pertain to the technology and the economics of the various systems. However, it should be stressed that many of the environmental, social and institutional problems, described, are also of great significance since, if not also resolved, they could cause the entire program of development and implementation, of the various types of solar energy systems, to be delayed or perhaps eventually abandoned.

Systems Implementation

In both of the scenarios (Accelerated Production and Minimum Production) for implementation of heating and cooling of buildings that follow it is assumed that the number of buildings in the future will be keyed to growth in population, as well as growth in commerce, industries, etc. Consequently, it is assumed that:

> Construction rates for new buildings will be maintained at a rate of 2.3% per year throughout the period 1975 to 2020.

> For housing, the total number of units will grow from 67 million in 1975 to 100 million units in 2020. Of these, about 60% will be single-dwelling units and about 40% multi-dwelling units.

> For all types of buildings, some new construction will be used to replace old buildings, and some will be used to provide additional occupancy for growing population, commerce, industries, etc.

> By 1985, 1.8×10^{16} Btu/yr, or 15% of the total energy consumption in the U.S., will be used for heating and cooling of buildings. As the number of buildings increases, their insulation qualities improve, and the power capacity per capita grows, the power requirements for heating and cooling of buildings will increase, but the percentage of the total energy consumption used for this purpose will vary, i.e., by the year 2000 2.0×10^{16} Btu/yr, or 11%, will be used for heating and cooling of buildings, and in the year 2020 2.3×10^{16} Btu/yr, or 13%, will be used for this purpose.

> The solar units will supply, on the average, two-thirds of the energy requirements for heating and cooling, and water-heating of buildings.

> Solar units will have an average lifetime of 20 years, and production of solar units will be increased, at an appropriate time, to allow replacement of discarded units.

> For the Accelerated Production Program, the goals are to equip 2% of all buildings with solar space-heating and -cooling, and water-heating units, by 1985 (this should be sufficient to satisfy about 0.2% of total U.S. energy requirements by 1985); 27%

(i.e., 2% of the U.S. total energy requirements) by 2000; and 60% (i.e., 5% of the U.S. total energy requirements) by 2020.

For the Minimum Production Program, the goals are to equip 0.7% of all buildings with heating and cooling and water-heating units by 1985. This should be sufficient to satisfy about 0.07% of the total U.S. energy requirements by 1985; 9% of the buildings (i.e., 0.7% of the U.S. total energy requirements) by 2000; and 23% of the buildings (i.e., 2% of the U.S. total energy requirements) by 2020.

Accelerated Production Program: Table 5.3 presents a typical schedule for producing solar heating and cooling, and water-heating equipment for the housing sector of the building industry, under an Accelerated Production Program. This assumes that the production of solar units will start in 1980, after the completion of the POCEs for solar space-heating and -cooling, and water-heating.

The initial production is assumed to be sufficient to equip 50,000 housing units per year, in 1980, and will rise to a level sufficient to equip a maximum of 1.9 million housing units per year, by the year 2001. It is assumed that this maximum production rate will be maintained (together with additional production required for replacement units) through the year 2020. This will result in a total of 1.5 million housing units equipped with solar space-heating and -cooling, and water-heating facilities by 1985; 23 million by 2000; and 61 million by 2020.

The above type of schedule, if applied to all buildings, would be expected to achieve the goals, stated above, for the Accelerated Production Program.

Minimum Production Program: Table 5.4 presents a typical production schedule for producing equipment under a Minimum Production Program. This, too, assumes that production of solar units will start in 1980, after the completion of the POCEs for solar space heating and cooling, and water-heating.

In the example of this scenario it is assumed that the initial production will be only sufficient to equip 20,000 housing units per year in 1980, and will rise to a level sufficient to equip a maximum of only 0.76 million housing units per year, by the year 2000.

In the example of this scenario it is assumed that this maximum production rate of 0.76 million housing units per year, plus additional production required for replacement units, will be maintained until the year 2020. This will result in a total of 0.5 million housing units equipped with solar space-heating and -cooling, and water-heating facilities by 1985; 7.3 million by 2000; and 23 million by 2020.

The above type of schedule, if applied to all buildings, would be expected to achieve the goals, stated above, for the Minimum Production Program.

TABLE 5.3: ACCELERATED PRODUCTION SCHEDULE

YEAR	TOTAL NEW CONSTRUCTION	ADDITIONAL OCCUPIED UNITS	TOTAL OCCUPIED UNITS	ANNUAL SOLAR HOUSING ADDITIONS	TOTAL NO. SOLAR HOUSING UNITS	SOLAR HOUSING UNITS/ TOTAL UNITS
	Millions of Units	Millions of Units	Millions	Millions of Units	Millions of Units	Percent
1980	1.61 M	0.8 M	71.0 M	0.05 M		
1981	1.63	0.8	71.8	0.09		
1982	1.65	0.8	72.6	0.21		
1983	1.67	0.8	73.4	0.30		
1984	1.69	0.8	74.2	0.39		
1985	1.71	0.8	75.0	0.49	1.5 M	2%
1986	1.73	0.8	75.8	0.62		
1987	1.74	0.8	76.6	0.76		
1988	1.76	0.8	77.4	0.97		
1989	1.78	0.8	78.2	1.19		
1990	1.80	0.8	79.0	1.35		
1991	1.82	0.7	79.7	1.47		
1992	1.83	0.7	80.4	1.54		
1993	1.85	0.7	81.1	1.57		
1994	1.87	0.7	81.8	1.61		
1995	1.88	0.7	82.5	1.64		
1996	1.90	0.7	83.2	1.67		
1997	1.91	0.7	83.9	1.70		
1998	1.93	0.7	84.6	1.75		
1999	1.95	0.7	85.3	1.80		
2000	1.96	0.7	86.0	1.90	23 M	27%
2001	1.98	0.7	86.7	1.99		
2002	1.99	0.7	87.4	2.11		
2003	2.01	0.7	88.1	2.20		
2004	2.03	0.7	88.8	2.29		
2005	2.04	0.7	89.5	2.39		
2006	2.06	0.7	90.2	2.52		
2007	2.07	0.7	90.9	2.66		
2008	2.09	0.7	91.6	2.87		
2009	2.11	0.7	92.3	3.09		
2010	2.12	0.7	93.0	3.25		
2011	2.14	0.7	93.7	3.37		
2012	2.16	0.7	94.4	3.44		
2013	2.17	0.7	95.1	3.47		
2014	2.19	0.7	95.8	3.51		
2015	2.20	0.7	96.5	3.54		
2016	2.22	0.7	97.2	3.57		
2017	2.24	0.7	97.9	3.60		
2018	2.25	0.7	98.6	3.65		
2019	2.27	0.7	99.3	3.70		
2020	2.28	0.7	100.0	3.80	60 M	60%

Source: PB-231 142, December 1973

TABLE 5.4: MINIMUM PRODUCTION SCHEDULE

YEAR	TOTAL NEW CONSTRUCTION	ADDITIONAL OCCUPIED UNITS	TOTAL OCCUPIED UNITS	ANNUAL SOLAR HOUSING ADDITIONS	TOTAL NO. SOLAR HOUSING UNITS	SOLAR HOUSING UNITS/ TOTAL UNITS
	Millions of Units	Millions of Units	Millions	Millions of Units	Millions of Units	Percent
1980	1.61 M	0.8 M	71.0 M	0.02 M		
1981	1.63	0.8	71.8	0.03		
1982	1.65	0.8	72.6	0.07		
1983	1.67	0.8	73.4	0.10		
1984	1.69	0.8	74.2	0.13		
1985	1.71	0.8	75.0	0.17	0.5 M	0.7%
1986	1.73	0.8	75.8	0.21		
1987	1.74	0.8	76.6	0.24		
1988	1.76	0.8	77.4	0.28		
1989	1.78	0.8	78.2	0.32		
1990	1.80	0.8	79.0	0.36		
1991	1.82	0.7	79.7	0.40		
1992	1.83	0.7	80.4	0.44		
1993	1.85	0.7	81.1	0.48		
1994	1.87	0.7	81.8	0.52		
1995	1.88	0.7	82.5	0.56		
1996	1.90	0.7	83.2	0.61		
1997	1.91	0.7	83.9	0.65		
1998	1.93	0.7	84.6	0.69		
1999	1.95	0.7	85.3	0.74		
2000	1.96	0.7	86.0	0.76	7.8 M	9%
2001	1.98	0.7	86.7	0.79		
2002	1.99	0.7	87.4	0.83		
2003	2.01	0.7	88.1	0.86		
2004	2.03	0.7	88.8	0.89		
2005	2.04	0.7	89.5	0.93		
2006	2.06	0.7	90.2	0.97		
2007	2.07	0.7	90.9	1.00		
2008	2.09	0.7	91.6	1.04		
2009	2.11	0.7	92.3	1.08		
2010	2.12	0.7	93.0	1.12		
2011	2.14	0.7	93.7	1.16		
2012	2.16	0.7	94.4	1.20		
2013	2.17	0.7	95.1	1.24		
2014	2.19	0.7	95.8	1.29		
2015	2.20	0.7	96.5	1.32		
2016	2.22	0.7	97.2	1.37		
2017	2.24	0.7	97.9	1.41		
2018	2.25	0.7	98.6	1.45		
2019	2.27	0.7	99.3	1.50		
2020	2.28	0.7	100.0	1.54	23 M	23%

Source: PB-231 142, December 1973

Economic Viability Calculations

According to J.A. Eibling of Battelle, Columbus Laboratories, who prepared the economic viability calculations for the MITRE study, it is important to explain that the work leading to the estimation of the economic ratios was a cursory effort, limited to working out one specific example. The work was viewed as a first attempt which hopefully would give some idea of the relative costs of solar applications compared with conventional energy systems.

Equally important, however, it was hoped that the exercise would point up refinements needed in preparing this type of estimate and lead toward the establishment of a standardized procedure for making economic projections for solar energy systems. Indeed the exercise served to show the need for such a procedure to establish baselines for economic projections together with supporting data for research investigators to use in preparing their cost estimates. Among the items needed are data on the effects of scale factors and climate and parametric curves for the many variables introduced in economic studies including energy storage costs, collector costs, interest rates, fixed charges, geographical data, meteorological data, etc.

In approaching the exercise, certain basic assumptions were made. For example, breakthroughs were not relied on for any of the applications and conservative estimates on costs of solar equipment were taken. All equipment costs were based on 1970 dollars. Energy storage for 2 to 3 normal days of operation was provided. Other assumptions are given in the tables. In comparative studies, the costs of solar energy systems should be adjusted upward to account for the need to provide supplemental conventional energy at low load factors. This was not done in the example shown.

For the heating and cooling application the example of a house located in an average solar insolation region is taken. Most of the assumptions are given in Table 5.5, however some explanation is warranted here. A solar radiation level of 1,500 Btu/sq ft/day is considered representative of the radiation received on a tilted collector on a yearly average at about the 40th north parallel. At this latitude both heating and cooling are required and the cost of the collector can be spread over both functions. Nevertheless the economic ratio would increase slightly, i.e., in favor of solar energy, if examples for more southerly locations were used.

The costs of the collector and heat storage system are believed to be conservative and representative of what a business enterprise might be expected to encounter. Heat storage is provided for 3 days, necessitating supplemental conventional heating for approximately one-third of the heating requirements. It is assumed that all of the cooling requirements can be met with solar energy with 3 days of cold storage. For purposes of comparison, the conventional system uses natural gas for comfort heating and for service water heating and electric vapor compression for comfort cooling. This combination for conventional systems generally gives the lowest cost for much of the United States.

TABLE 5.5: HOUSE HEATING AND COOLING (Average U.S. Solar Insolation)

	1970		1985		2000		2020	
	Solar	Conv	Solar	Conv	Solar	Conv	Solar	Conv
Energy Required								
Heating, MBtu/yr	100 MBtu/yr	133 MBTU/yr	80	100	75	93	70	85
Cooling, MBtu/yr or Kwhr/yr	63 MBtu/yr	2930 Kwhr/yr	40	2090	32	1860	30	1760
SWH, MBtu/yr	20 MBtu/yr	33 MBtu/yr	16	23	15	21	14	19
Total, MBtu/yr or Kwhr/yr	183 MBtu/yr	166 MBTU/yr + 2930 Kwhr/yr	136	123 + 2090	122	114 + 1860	114	104 + 1760
Collector Data								
Solar overall eff. %	25		32		35		36	
Collector area, ft^2	958		620		520		460	
Unit Collector cost, $/ft^2	5.00		4.00		3.75		3.50	
Cost Equipment, $								
Collector	4800		2480		1950		1600	
In-house conventional	4000	4000	3300	3300	3200	3200	3000	3000
Extra cost solar equipment	1000		800		700		600	
Heat Storage	2000		1300		1100		1000	
Total	11800	4000	7880	3300	6950	3200	6200	3000
Annual fixed charges at 12%	1416	480	945	396	834	384	744	360
Cost Energy, $								
Heating and SWH	56	183	68	332	68	353	69	343
Cooling		62		53		51		52
Blower or pump	40	20	40	20	40	20	40	20
Total annual energy	96	265	108	405	108	424	109	415
Total Annual Cost, $	1512	745	1053	801	942	808	853	775
Economic Ratio, solar/ conv.	2.0		1.3		1.2		1.1	
Bases:								
Average solar on tilted collector, Btu/ft^2/ day	1500		1500		1500		1500	
Collector thermal eff., %	35		40		44		45	
Solar utilization eff., %	70		80		80		80	
Annual solar available, MBtu/ft^2	191		219		235		246	
Natural gas price, $/MBtu		1.10		2.70		3.10		3.30
Electricity price, c/Kwhr		2.10		2.52		2.73		2.94

Note: For conventional systems, heating is with natural gas and cooling is with electricity. For solar systems, cooling is with absorption unit. Energy requirements for heating and cooling arbitrarily reduced for 1985 and beyond to allow for more thermally efficient houses.

Source: PB-231 142, December 1973

Attention is called to the fact that the total demand for heating, cooling, and service water heating is arbitrarily decreased in subsequent time frames to allow for the fact that major emphasis will be placed on conserving energy by increasing the thermal efficiency of buildings. In 1970 the annual per family energy requirement is taken as 100 million Btu for comfort heating, 25 million Btu for cooling and 20 million Btu for water heating. The respective figures for the year 2020 are 70, 18, and 14. An attempt has been made to show all costs associated with house heating, cooling, and service water heating. The choice of 12% for fixed charges to include interest, depreciation, insurance and taxes is somewhat arbitrary. Some homeowners might regard their fixed costs as being lower than this, say 10%, in which case the lower percentage can be substituted in the table.

Because solar applications are extremely capital intensive, the main costs are associated with fixed charges on the solar collector and attendant energy storage systems. At present, high-quality flat plate collectors suitable for use in heating and cooling and water-heating cost about $5/sq ft of collector surface, installed. This cost results in an equivalent fuel cost of about three times present natural gas prices, but approximately equal to the costs of electric heating. By 1985, with gas prices predicted to be over $2\frac{1}{2}$ times the 1970 price, the heating and cooling of buildings with solar energy is expected to be near the costs of conventional systems.

If the cost of the collector can be reduced to $3.50/sq ft through mass production and improved construction techniques, while concurrently improving the collection efficiency through the use of better selective coatings, it is possible that by the year 2000 heating and cooling with solar energy will cost even less than with conventional fuels. It is estimated that approximately one-third of the U.S. requirement for building heating and cooling could be met by the year 2000 if a major effort is made to equip all new buildings with solar systems and to provide retrofit systems for many existing buildings.

General Electric Study

BACKGROUND

The material in this chapter is from NSF-RA-N-74-021A, *Solar Heating and Cooling of Buildings, Phase 0—Feasibility and Planning Study—Final Report,* an executive summary completed by General Electric Company for the National Science Foundation in May 1974. Portions of this work were done by the University of Pennsylvania and the Ballinger Company. Supplementary data used in the preparation of this study are available as PB-235 432 (NSF-RA-N-74-021B), PB-235 433 (NSF-RA-N-74-021C), PB-235 434 (NSF-RA-N-74-021D) and PB-235 435 (NSF-RA-N-74-021E).

Studies on Solar Heating and Cooling of Buildings (SHACOB) were initiated in September 1973 under the direction of the Research Applied to National Needs (RANN) branch of the National Science Foundation (NSF) to provide baseline information for the widespread application of solar energy. The potential application is dependent upon many factors—costs, systems performance, climatic variation, societal, legal and environmental aspects are representative of these.

The solution of such varied, interrelated problems requires a systematic, multi-disciplinary approach. Such an approach was structured by NSF. It involves participation by government, universities, and industry. The objective of the initial phases of this program is to set the stage for widespread utilization by establishment of the basic feasibility (Phase 0) and the design (Phase 1) and implementation (Phase 2) of Proof-of-Concept Experiments (POCE).

One major objective of Phase 0 is the establishment of an overall experimental program plan to provide visible evidence of the practicality of heating and cooling buildings with solar energy within the socioeconomic environment of the United States.

SCENARIO PROJECTIONS

This project is concerned with a time frame of at least 25 years, from now until AD 2000. Assumptions on the expected or projected state of society over this period will constitute a baseline scenario. This scenario, significant elements of which are shown in Figure 6.1 postulates no major wars or depressions during the period to AD 2000, although it is hard to find a twenty-five year period in the country's history for which these postulates are true. If these happen, the alternative scenarios they produce are so varied that no reasonable surprise-free scenario can include them. All dollar values used in this study are given in terms of 1970 dollars.

A sound starting place for economic assumptions about the period to AD 2000 is a projection of the future population and the future productivity per worker. While other factors enter, these are the two most important determinants of the quantitative economic growth.

In the baseline scenario of Figure 6.1a the population of the United States will grow from 205 million in 1970 to 264 million in 2000, an average growth rate of 0.85%. This corresponds to the Bureau of the Census projection known as Series E (Bureau of Census, 1972). The baseline projection GNP in AD 2000 will be $2,400 billion compared to $977 billion in 1970. This is considerably lower than many estimates made within the past few years.

Economic indicators derive from the total labor force and estimates of the production per worker and the unemployment rate. The GNP is divided into categories called kinds of purchases. Personal consumption and government expenditures will grow most rapidly. Net exports may vary widely from year to year.

The kinds of outputs which these purchases buy are categorized as industries. Government will grow the fastest at an average rate of 3.7%. Both contract construction and services will grow faster than the 3% rate for GNP. Manufacturing, agriculture and mining will grow slower than the GNP. These rates symbolize the state of society: well beyond the agriculture stage, and passing rapidly from the industrial state to a postindustrial era.

Alternate high and low scenarios have been considered to provide the range of uncertainty. The high and low estimates of total population in AD 2000 are 300 million and 251 million. The high estimate of GNP in AD 2000 is $3,400 billion and assumes an average growth in productivity (dollars of GNP per worker) of 2.75%, while baseline assumes average growth of 1.5%.

One of the major considerations affecting economic viability of solar energy systems is the price of the fuel whose consumption is saved. Figures 6.1b and 6.1c present the projected retail fuel prices for various energy sources. For dollar energy assessments, the retail fuel price at the residence or commercial level is the the most significant. The projection for retail prices of fuel into the future is

FIGURE 6.1: ECONOMIC AND ENERGY SCENARIOS

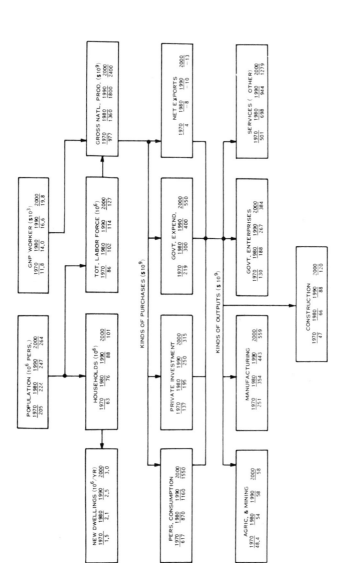

a.

(continued)

FIGURE 6.1: (continued)

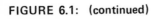

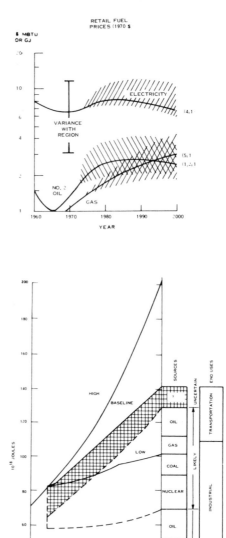

highly speculative since it involves balancing many contradicting factors. The fuel price projections were arrived at by a detailed examination of the interaction of supply and demand, based on historical precedents and projections of technological, economic and political developments. The units are expressed in terms of fuel energy content and in 1970 dollars. The solid curves are the baseline values used to assess solar systems cost effectiveness.

Fuel shortage in the face of increasing demand, coupled with uncertainty of future supplies, is another driving force to develop alternate energy supplies. As shown in Figure 6.1. a significant gap exists between readily available supplies and the growing needs of the country.

SYSTEMS DESCRIPTION

The basic elements of a solar heating/cooling system are depicted in Figure 6.2. Generally, the system may be considered as one which incorporates solar energy equipment with conventional heating, ventilating, and air conditioning components. A description of the various subsystems follows.

The heart of the system is the solar collector. These are being designed and tested but the industrial base to provide economic, reliable equipment does not exist today. Achievement of maximum energy savings on a cost-effective basis also requires specially designed air conditioning with improved characteristics.

The heat collector subsystem consists of the solar energy heat collection unit such as flat plate solar collectors, water encapsulation devices, or thermal panes. This is the heart of the system, and accounts for approximately half the cost. Development of more efficient, less expensive collectors will be a continuing need for solar energy to reach its full potential.

The thermal energy storage subsystem provides for the storage of solar energy during periods of surplus of sunshine for use at night or on cloudy days. Storage methods include sensible and latent heat transfer (hot or cold) components, including combinations and/or composites of the two. Here again this specialized solar energy equipment requires improvements, primarily for reducing costs of large installations.

The environmental conditioning subsystem includes all ancillary components and equipment required to convert and distribute energy to condition the space air within the building. It includes all circulating pumps, valves, condensers, cooling equipment, controls, etc., and supplementary equipment required to condition the space air by gas, oil, or electrical energy.

The energy management and control subsystem provides the necessary control system logic and components necessary to sense and activate system controls.

FIGURE 6.2: ELEMENTS OF SHACOB SYSTEMS

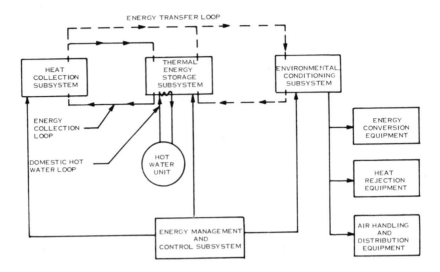

TECHNICAL EVALUATIONS

The functional requirements for heating and cooling systems are primarily influenced by climate and type of building. While local variations in climate occur
everywhere, the United States, excluding Alaska and Hawaii, may be divided
into twelve climatological regions as shown in Figure 6.3.

FIGURE 6.3: CLIMATIC REGIONS IN THE UNITED STATES

In this study each region is designated by a city where climate is representative of the region. Broad areas of the country are therefore covered by each city designator.

Building types are extremely varied, but they can be categorized for purposes of estimating heating and cooling loads. For this study all buildings which are candidates for solar energy installations are divided into eighteen categories as shown in Table 6.1.

TABLE 6.1: BUILDING CATEGORIES

    (1) Residence, one and two family

    (2) Residence, multiple high rise

    (3) Residence, multiple low rise

    (4) Hotel/motel, high rise

    (5) Hotel/motel, low rise

    (6) Office building, high rise

    (7) Office building, low rise

    (8) Warehouse

    (9) Industrial, light process load

   (10) Industrial, heavy process load

   (11) Single story educational

   (12) College/university

   (13) Auditoriums

   (14) Health care, clinic

   (15) Hospital

   (16) Retail, merchandise mall

   (17) Retail, individual store

   (18) Mobile homes

In addition to the climate and building combinations, there are many alternate solar energy system concepts for heating and cooling. The main thrust of the technical/cost evaluation was to identify those solar system designs with the greatest potential for life cycle cost savings and estimate the annual energy savings which would accrue from the application of each selected solar system.

Capital and operating costs were derived through analysis of HVAC (heating, ventilating and air conditioning) design, installation and maintenance practices, standard component prices, historical cost trend data, estimated costs for specialized solar equipment, and fuel projections.

The life cycle cost for such systems studied considered such factors as component, integration and installation costs; appropriate building types and climate areas; year of installation, useful life, capital and operating costs; and interest rates and tax savings on capitalization.

This approach presents all costs on a comparable basis without the additional uncertainty of superimposed inflationary effects. Cost sensitivity analyses were made by varying the baselines for such significant factors as fuel and solar collector costs, interest rates and capitalization periods.

The solar collectors represent a major element in system cost (typically around 50%). Since they are not currently available commercially, their prices were estimated on the basis of projected designs produced in a plant with a capacity of approximately one million 4 ft by 8 ft panels per year. It was assumed they would be sold, distributed, and installed like other major HVAC components which have a sale price to the final purchaser between 2.4 and 2.7 times the base manufacturing cost.

On the above basis, user costs are projected to be $4.70/ft^2 for single-window units and $5.80/ft^2 for double windows. These estimates are in 1974 dollars, but the cost-effectiveness computer calculations adjusted them to be consistent for all cases considered. These estimates are considered realistic for the initial period of significant commercial application. It is highly probable that they will be reduced if a continued R&D program for reducing collector costs is adopted.

Collector performance is based on analyses which have not yet been verified experimentally, particularly for large arrays. The tentative conclusions are that double-window collectors are more cost effective in northern climates and that single-window collectors are more cost effective for southern climates.

With the complex matrix of solar systems, building types, and climate zones, it would be hazardous to draw oversimplified conclusions regarding preferred systems, especially with the varied state of technical development of such components as solar collectors and air conditioning equipment designed for use with solar systems.

Four systems were found to be especially worthy of further evaluation, development and demonstration:

Heating with liquid-to-air heat exchangers coupled with absorption air conditioners modified for solar operation. Figure 6.4 is a schematic diagram of this concept.

Heating with liquid-to-air heat exchangers coupled with conventional vapor compression air conditioners.

Heat pumps assisted by solar energy in the heating mode and operated either conventionally or in an off-peak manner (stored cold) for cooling.

Heating with liquid-to-air heat exchangers and cooling with vapor compression air conditioners driven by Rankine cycle engines designed for operation with solar energy.

Under certain conditions, each of these systems can be cost effective on a life-cycle basis. It is important to note further that even when cost effectiveness is marginal, substantial energy could be saved. For instance an important design variable considered is the ratio of solar collector area to roof area. It is generally true that energy savings increase as this ratio increases.

However, this is not necessarily true of cost effectiveness, since the added collector cost may exceed the value of the fuel saved at projected prices. The relative value of these competing standards could be influenced by Government policy actions. The quantitative data for evaluating these alternatives is presented in detail in the report.

The evaluation of each building type in each climate area with the various solar systems yielded a ratio of life cycle cost with solar to conventional systems for each combination. When this ratio is 1.15 or less, the solar system is assumed to be cost effective. This practice was adopted to make some allowance for improvements in performance and cost as R&D progresses. It was also assumed that the dominant trend for future conventional systems will be all electrical. This assumption is valid for the long range projections of interest.

Cost effectiveness of solar cooling systems was based on performances characteristics calculated for present air conditioner designs. On that basis cooling plays only a small role in determining the monetary savings of solar energy systems. One major conclusion of the study is that cost effective solar air conditioning depends upon the development of equipment optimized for this purpose, rather than current equipment designed to use conventional energy resources.

The issue of installing solar systems on existing buildings was examined for a number of building types. Aesthetic and structural requirements tend to make these costs both higher and more variable. For those generic types which can be modified satisfactorily in other respects, the costs were found to be so much higher that they would not be cost effective. There are, of course, many individual buildings fortuitously adaptable, but their number is small compared to the major market in new construction.

In Tables 6.2 and 6.3 following, summary data is given for the number of buildings and total energy savings possible in the year 2000, if SHACOB systems were installed in all suitable buildings. This number of buildings is referred to as the capture potential and represents the ultimate market.

FIGURE 6.4: TYPICAL SOLAR HEATING AND COOLING SYSTEM SCHEMATIC

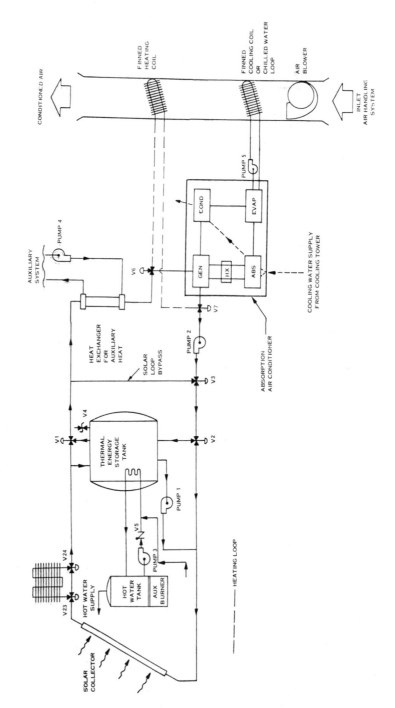

TABLE 6.2: CAPTURE POTENTIAL AD 2000—NUMBERS OF BUILDINGS*

	Climate Area									
	Wash., D. C.	Boston	Madison	Bismarck	Los Angeles	Phoenix	Ft. Worth	Omaha	Charleston	Total
Residence, One & Two Family	4,133	8,237	3,075	1,371	3,656	540	1,826	1,308	3,190	27,336
Residence, Multiple Low Rise	444	1,383	302	124	475	39	146	126	239	3,278
Hotel/Motel, Low Rise	10	31	10	5	12	5	7	4	8	92
Office Guildings, Low Rise	118	332	165	64	96	21	46	52	107	1001
Warehouse	69	221	82	28	60	11	29	23	47	570
Industrial, Light Process Load	59	278	75	38	162	10	24	31	36	713
Education, Single Story	9	27	8	4	13	1	5	4	10	81
College/University	2	5	2	1	4	1	3	1	4	23
Auditoriums	54	125	44	16	73	8	24	21	43	408
Retail, Merchandise Mall	2	6	2	1	3	-	2	1	3	20
Retail, Individual Store	222	622	197	81	233	34	95	70	159	1,713
Mobile Homes	554	714	473	265	519	224	234	226	669	3,878
Total	5676	11981	4435	1998	5306	894	2441	1867	4515	39,113

*In thousands.

TABLE 6.3: CAPTURE POTENTIAL AD 2000—ENERGY SAVED*

Building Types	Washington D. C.	Boston	Madison	Bismarck	Los Angeles	Phoenix	Fort Worth	Omaha	Charleston	Total
Residence										
One and Two Family	244	527	191	144	201	25	110	76	179	1697
Multiple Low Rise	23	80	16	6	12	1	5	6	7	156
Hotel/Motel, Low Rise	28	77	19	17	8	6	5	8	5	173
Office Buildings, Low Rise	42	111	40	15	15	2	7	13	15	260
Warehouse	97	260	71	56	81	7	19	26	40	657
Industrial, Light Process Load	136	452	104	36	157	8	29	36	43	1001
Education, Single Story	15	41	12	5	12	1	3	6	8	103
College/University	2	7	2	1	1	1	1	1	2	18
Auditoriums	16	39	11	8	10	1	4	5	7	101
Retail										
Merchandise Mall	8	29	9	8	2	1	3	4	4	68
Individual Store	121	315	99	40	42	5	28	37	30	717
Mobile Homes	17	23	15	14	15	5	7	6	18	120
Total	749	1961	589	350	556	63	221	221	358	5071

*In millions of gigajoules.

CAPTURE POTENTIAL AND MARKET PENETRATION PROJECTIONS

An assessment of the capture potential and market penetration for solar heating and cooling by building type and climate area has been projected through the year 2000. This assessment includes forecasts of energy savings attributable to solar heating/cooling installations.

Capture potential, the number of buildings that could be equipped effectively with solar systems, is less than the total number of occupied buildings. This is attributable to considerations involving climate areas, building types, siting, viability of solar cooling systems, retrofit vs new installations, and economic viability.

The projections for numbers of buildings of each type to be constructed in each climate region were based on detailed economic and demographic forecasts. For this purpose the United States was divided into 171 Business Economic Areas (BEA), and individual forecasts were made for each. Of the 60 million buildings to be constructed in the United States in the next 5 years, approximately 40 million were found to be viable, cost effective, candidates for solar systems.

The number of buildings suitable for solar energy systems from 1975 to 2000 is indicated in Figure 6.5a. If all these buildings were so equipped, the yearly equivalent electric power savings would be approximately 1,500 billion kwh (Figure 6.5b) by the end of the century, equivalent to the total electrical generating capacity of the United States in 1970. Also, as indicated by a comparison of Figures 6.5c and 6.5d the annual value of the solar savings of fuel would, by that time, exceed the annual additional outlay for the solar systems.

The annual costs in Figure 6.5d are the increment attributed to the solar systems. In many building types, substantial savings in roofing material can be made by substitution of solar collectors. Since this factor is taken into account, the cost of the solar equipment itself is higher.

While the capture potential represents an ultimate goal, its achievement by the year 2000 is limited by a number of factors. The annual cost of installing solar equipment in that time period would be of the order of 400 billion dollars. This estimate is high because it does not account for all the potential cost reductions which could be obtained by future R&D.

The total, however, is still much larger than the market which can realistically be served in the time period considered. Factors limiting penetration of the potential market include commitment of risk capital, development and emplacement of manufacturing capacity, training personnel and overcoming public inertia to new products.

A major conclusion of this study is that the growth of the market for solar energy systems during the 1980's can be impacted more by early industrial participation than by the availability of cost-effective applications.

FIGURE 6.5:   CAPTURE POTENTIAL PROJECTION SUMMARY (NEW
CONSTRUCTION ONLY)

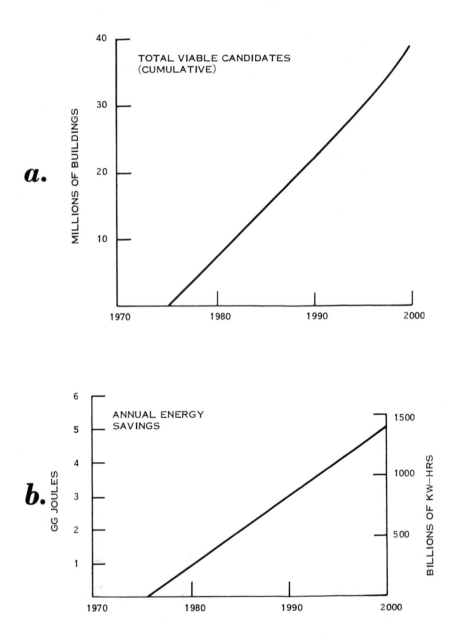

(continued)

FIGURE 6.5: (continued)

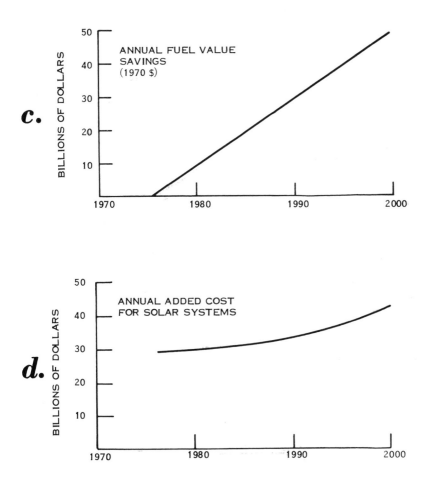

The projected market penetration, summarized in Figure 6.6, is based on historical data for similar industries, augmented by estimated effects of government
policies designed to stimulate the use of solar energy. Policy alternatives were
explored in depth in the study. The major ones considered most likely to be
effective include:

Proof-of-concept experiments to provide improved design tools, operating experience, equipment and visibility to early decision makers.

An early introduction of installations on government buildings, new and
retrofit, which will provide an initial market to speed the development
of the industrial base for equipment, distribution and service.

Economic incentives, such as loan guarantees and interest incentives, to private purchasers to tip their decision balance in favor of the favorable solar system life-cycle costs.

Establishing the requisite legal safeguards (such as sun rights) and minimizing zoning restrictions.

Continued sponsorship of R&D to assure that economic, reliable equipment and systems will be available as required.

The projections summarized in Figure 6.6 result from the baseline scenario for fuel costs, solar equipment costs and other factors. Higher fuel costs, reduced per unit solar equipment costs, and higher efficiency would yield higher market penetration and fuel savings. The projections indicate adoption of solar heating and cooling at an accelerating rate as we enter the twenty-first century. The in-depth detailed analyses were made only to the year 2000, but less exhaustive analysis for the subsequent period indicates a far greater impact by 2020.

FIGURE 6.6:   MARKET PENETRATION PROJECTION SUMMARY (NEW CONSTRUCTION ONLY)

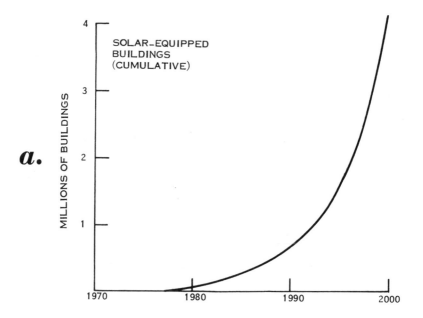

(continued)

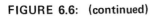

FIGURE 6.6: (continued)

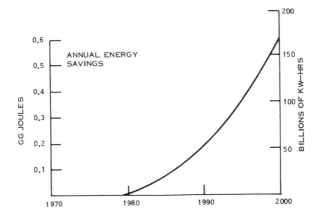

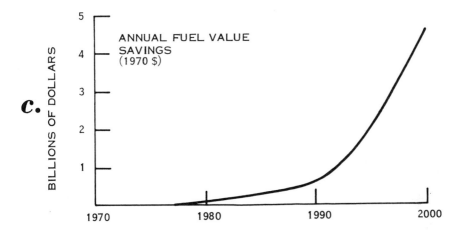

(continued)

FIGURE 6.6:　(continued)

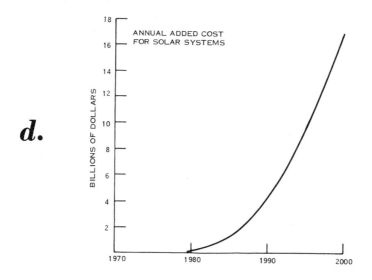

d.

PROOF-OF-CONCEPT EXPERIMENTS (POCE)

Achieving the projected potential for solar energy requires a number of early initiatives. Among the most important are a series of proof-of-concept experiments to assess true viability and acquire the data needed for further major commitments. These experiments should be emplaced in locations representative of the climate region of the United States. They should also cover a representative spectrum of building types and solar energy system concepts.

Final selection of the specifics of individual sites and systems must await the program implementation phase. A detailed analysis was made in the study, however, to establish criteria, determine locations (climate areas and building types), number and type of experiments, work flow schedules, and management structure. The most important criteria in selection of specific experiments are:

Representativeness of projected national energy consumption patterns and the degree to which the data can be generalized, considering both current and expected future designs.

Ability to provide a substantial portion of the heating and/or cooling load (at least 30%).

Ability to obtain performance, reliability, cost and operating data.

Visibility of the installation to decision makers in such areas as architecture, building and equipment specification, building financing and public policy.

Degree to which consuming public awareness of solar energy will be enhanced.

The first two criteria apply to selection of climate areas, buildings types and experiment category (heating, cooling, or both). The acquisition of the data to assess these factors was discussed in the sections on systems evaluation and capture potential. An analysis of those data led to the recommendation of 82 experiments as shown in Table 6.4 to assess viability and lay the basis for expanded commitments. They include heating and cooling in each climate area on those buildings having the largest impact on national energy consumption.

TABLE 6.4: PROOF-OF-CONCEPT EXPERIMENT RECOMMENDATIONS

No.	Building Type — Description	Mad.	Chars.	L.A.	Bos.	D.C.	Phoen.	Miami	Omaha	Bism.	Nash.	Seattle	Ft. W.
1	Residential, Single Family	4H	2HC	3H	6H 1HC	3H	1HC	3C	1CH	1H	2HC		3HC
2	Residential, Multiple High Rise												
3.	Residential, Multiple Low Rise												
4	Hotel/Motel, High Rise												
5	Hotel/Motel, Low Rise						1C						
6	Office Building, High Rise												
7	Office Building, Low Rise	2H	3HC	1H 1HC	5H 1HC	3HC	1HC		1H	1H	2HC	1H	2HC
8	Warehouse												
9	Industrial, Light Process Load	1H		2H	6H	2H			1H	1H			
10	Industrial, Heavy Process Load												
11	Educational, K to 12												
12	College/University												
13	Auditoriums	2H		1H	5H 1HC	2H					1H	1HC	1HC
14	Health Care, Clinic												
15	Hospital												
16	Retail, Merchandise Mall												
17	Retail, Individual Store												
18	Mobile Home												

H = Heating only
C = Cooling only
HC = Both Heating and Cooling

An overlapping two-phase program was studied for implementing this recommendation. Under this plan all installations would be in place within 24 months of program initiation. Site selections experiment designs and other activities preceding construction would be completed in 7 months for early experiments, and in 15 months for the entire array. In general, retrofit and heating-only experiments would be in the early group. New construction and cooling would follow. The construction periods would vary from a few months for some retrofit heating units to approximately a year for some new buildings.

Achieving maximum benefit from the POCE program will require careful and skilled management. The report outlines a management plan based on a prime contractor handling blocks of experiments and utilizing the services of architect/engineers who, in turn, would employ general contractors. In each case, the responsible building official (landlord) would be involved during selection, installation and operation. A program systems engineering and integration activity is necessary to assure consistency in data acquisition, analysis and interpretation. Such a program, properly executed and evaluated, would be a springboard for the broader applications of solar energy.

PROGRAM IMPLEMENTATION PLANS

The purpose of these plans is to recommend an orderly route by which SHACOB can progress from its present status of component experiments and limited installations into widespread utilization. The study has indicated that technical feasibility and economic viability are both achievable, but not instantaneously nor easily. Because of a national need to conserve conventional energy sources, it is appropriate that the federal government provide for the seeding of SHACOB via installation in government buildings and sponsorship of related R&D programs. A schedule of distinct steps to implement the utilization of solar heating and cooling is shown in Figure 6.7.

FIGURE 6.7: PRELIMINARY UTILIZATION PLAN

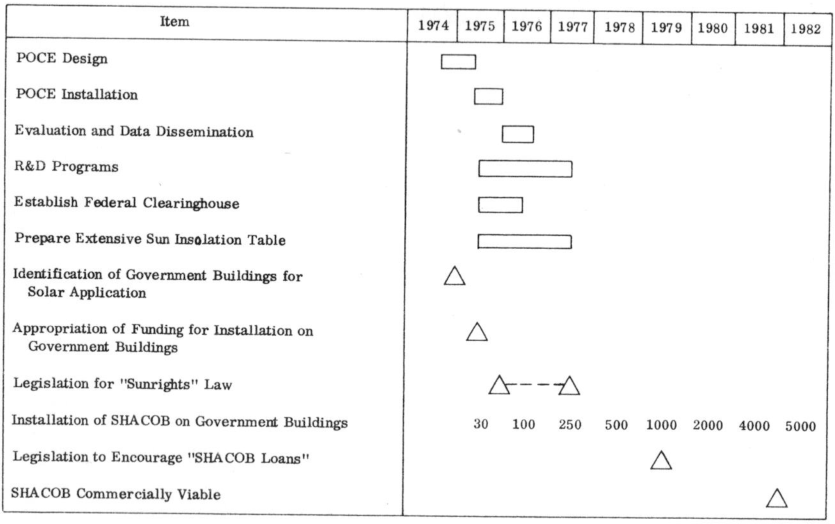

Item	1974	1975	1976	1977	1978	1979	1980	1981	1982
POCE Design		▭							
POCE Installation		▭							
Evaluation and Data Dissemination		▭							
R&D Programs		▭▭▭							
Establish Federal Clearinghouse		▭							
Prepare Extensive Sun Insolation Table		▭▭							
Identification of Government Buildings for Solar Application	△								
Appropriation of Funding for Installation on Government Buildings	△								
Legislation for "Sunrights" Law		△---△							
Installation of SHACOB on Government Buildings	30	100	250	500	1000	2000	4000	5000	
Legislation to Encourage "SHACOB Loans"						△			
SHACOB Commercially Viable								△	

Commencing with the Proof-of-Concept-Experiment, is the preparation of a time-phase communications plan. This plan is geared to disseminate information, at all stages of the project, to appropriate key actors in the selection of heating and cooling systems and to the general public. A federal clearinghouse for data on all aspects of solar heating and cooling should be established. An important element is the opening of a real market for SHACOB equipment. This will encourage industry to perform the product engineering and to put into place the tooling, facilities, and people to fabricate, install, and maintain solar equipment. A commitment by the government to install solar equipment on a significant number of buildings over the next ten years could provide the basis for private sector installation in the 1980's.

IMPACTS OF SOLAR HEATING AND COOLING

The implementation of SHACOB at the rate projected by the market penetration will have an impact on the environment, economy, industry, utilities, government policy, law, and society.

Environment

The primary environmental effects of SHACOB are the reduced conventional fuels required, the consequent reduction in land use and pollution associated with the conventional energy production and the increased consumption of materials used in manufacturing solar energy collectors and components. The energy savings in AD 2000 represent 20 electric generating plants, each with a capacity of 1,000 megawatts operating at 90% load factor.

The pollution impacts are impressive: 390 million kilograms (430 thousand tons) of air emissions, 105 million kilograms (115 thousand tons) of water discharges, and 18 billion kilograms (20 million tons) of solid waste avoided; and 1 million curies in air and water discharge plus 670 million curies of solid waste not generated. Consumption of material resources in the manufacture of solar energy collectors is significant but not excessive.

Economic

The primary economic impact is the increased capital investment requirement it imposes on the cost of buildings. The development of a 10 billion dollar a year solar collector and component industry by AD 2000 would contribute to economic growth through construction of new factories, increased employment in manufacture and installation and an improved trade balance due to reduced dependence on fuel imports.

The impact on conventional energy utilities resulting from decreased sales coupled with the necessity of serving a growing number of customers is difficult to quantify. The market penetration of SHACOB could have an effect on land values. Land with good southern exposure in high insolation regions would be enhanced in value.

Utilities

SHACOB systems are most effective in providing energy at the time of day when the electric utilities experience their peak load. The forecast of off loading of only 2% of the energy demand by AD 2000 indicates that on a national scale the impact will be relatively small.

Policy and Legal

Because of the long period which must elapse before SHACOB can make a significant impact on the U.S. energy picture, the eventual effect of SHACOB as a nonpolluting source is its most important characteristic. The forcing function for SHACOB is environmental policy rather than energy policy. Solar energy will look more toward political mechanisms than toward market mechanisms.

The technology will not be implemented primarily in response to consumer sovereignty but because it may be prescribed for use by government and in buildings that are in part federally financed. Federal laboratories such as the National Bureau of Standards will have an increased role in setting standards for SHACOB systems and Federal regulatory agencies will monitor conformity to their use.

Changes in law may be necessary to treat solar radiation as a common property resource. Legislative action to resolve questions of sun rights and the authority of local governments to engage in three dimensional zoning will be necessary.

Social

The newly emergent social ethic oriented toward quality of life rather than quantity of goods may become a fundamental tenet of society by AD 2000. Continued movement of the population to the South and Southwest will facilitate introduction of novel architectural forms adapted to use of SHACOB technology. The use of solar energy would tend to change the use of vegetation as an aid to interior comfort conditioning.

CONCLUSIONS

Heating and cooling with solar energy can replace large quantities of fuel. The attendant reduction in pollution will attain an ever-increasing significance as total energy consumption increases. The industrial base and public acceptance required for achieving these beneficial results must be developed over a period of time with a planned program involving both the public and private sectors.

To use solar energy for widespread heating and cooling in the future requires that the nation begin to develop the tools, the techniques, and the industries today. Early Proof of Concept Experiments are an excellent starting place for a number of reasons. They will provide:

Early design and operational experience.

Actual performance and cost data for a variety of climate zones and building types.

Comparative performance data on solar energy system concepts and equipment.

Identification of the problem areas and solutions for the next generation of equipment.

Education tools for the building industry.

After the Proof of Concept Experiments, the next phase should be wide utilization on governmental buildings, both new buildings and retrofit. This will achieve:

An initial market to establish industrial incentive to supply the products and services.

Distinct, measurable savings of conventional fuel supplies.

Innovations and improvements based on these applications.

By the end of the century the projected capture potential for cost effective solar energy installations is approximately 40 million buildings. If all were solar equipped, the annual energy savings would exceed 1,500 billion kilowatt-hours. Cost effective capture potential is dominated by new construction.

Market penetration will be limited primarily by the development of the industrial base. With implementation of effective government policies, it will be established by the mid-eighties and grow rapidly thereafter reaching annual energy savings of the order of 150 billion kilowatt-hours by the year 2000. Its growth rate in that time period will yield a far greater impact by the year 2020. The realization of the projected market penetration by the year 2000 would

Reduce annual air emissions by over 430 thousand tons.

Avoid about 20 million tons of solid waste annually.

Reduce radioactive air and water discharge by over 1 million curies and solid waste by over 670 million curies.

Introduction of cost effective solar air conditioning equipment would substantially increase the projected capture potential and market penetration in terms of energy savings. The most important government policies needed to stimulate solar heating and cooling are in the following areas.

Proof of concept experiment implementation.

Early introduction of solar systems on government buildings (new and retrofit).

Economic incentives to encourage a life-cycle approach to HVAC systems.

Legal safeguards and minimizing zoning restrictions.

Sponsorship of R&D for economic, reliable equipment.

The General Electric Company, in cooperation with NSF, has recently designed, installed and is currently operating a major experimental installation at the Grover Cleveland Junior High School in Boston.

Westinghouse Study

BACKGROUND

The information in this chapter is adapted from NSF-RA-N-74-023A, *Solar Heating and Cooling of Buildings—Executive Summary,* prepared by Westinghouse Electric Corporation for the National Science Foundation, issued May 1974. Also contributing to this study were Colorado State University; Carnegie-Mellon University; Burt, Hill and Associates (architects); NAHB (National Association of Home Builders) Research Foundation, Inc. and PPG Industries, Inc. Supplementary information for this study is available as PB-235 427 (NSF-RA-N-74-023B), PB-235 428 (NSF-RA-N-74-023C) and PB-235 429 (NSF-RA-N-74-023D).

Innovations in building designs are adopted slowly. Since single-family residences represent the largest investment that a family normally makes, it is understandable that a conservative clinging to tried techniques and designs tends to prevail. Building innovations that increase initial costs are also normally introduced at the narrow top of the economic ladder or in commercial buildings first, and trickle down to a broader base over a protracted period of time.

Central air conditioning for homes is a pertinent example of these conclusions. It was available in the early 1930's but incorporated in only a relatively few high-priced homes. Not until the 1960's did costs decrease sufficiently and middle-income-family aspirations rise to create a volume demand for central air conditioning. Today, in many sections of the country, few homes are built without it. Residential heating and air conditioning systems are considered to have a useful life of about 15 years. Therefore, the solar and conventional systems considered in this study have been analyzed on a 15-year-life-cycle basis.

Feasibility of solar systems is strongly dependent on present and future costs of equipment and conventional forms of energy. Solar system equipment costs can be expected to come down as technology improves and production volume increases. Fossil fuel prices can be expected to increase. In this analysis, future costs are projected on a normal, rational growth basis. However, sudden changes can occur, particularly in gas and oil prices, that could create a more urgent demand for solar systems.

TECHNICAL CONSIDERATIONS

Requirements

The best opportunities for the greatest savings of energy are to be found in those situations where most of the energy is consumed. Consumption of energy for space heating and cooling is determined by population and climate. Population establishes the requirement for buildings, while the climate determines the heating and cooling characteristics of buildings and the heating and cooling load. Thus, an analysis of demographic and climatic relationships would suggest the best opportunities for energy savings.

Energy Consumption

Table 7.1 summarizes the annual energy consumption for space heating, space cooling, and domestic hot water in the United States, and compares it to the total energy demands (including transportation, power, etc.) of the Nation. It is significant that heating and cooling requirements of buildings constitute more than a quarter of the total annual energy consumption of the Nation. It is not surprising that most of the energy for space heating is used in the North while that for space cooling is used in the South, and hot water demand is fairly uniform throughout the United States. The table shows that space heating dominates the demand for energy now and that space cooling demand is expected to rise sharply.

TABLE 7.1: ANNUAL ENERGY CONSUMPTION OF THE UNITED STATES

	1968		1980	
	10^{12} Btu	Percent of Total	10^{12} Btu	Percent of Total
Space Heat	12,105	20.0	19,161	18.4
Hot Water	2,416	4.0	4,110	3.9
Space Cooling	1,540	2.5	5,320	5.2
Subtotal	15,961	26.5	28,591	27.5
National Total	60,529	100.0	104,000	100.0

Regions

A demographic analysis identified the regions of the country where the population densities are greatest. Taken together with the climatic conditions that determine the type of demand for heating and cooling systems, these areas would constitute the principal potential commercial market for solar systems. The remainder of the area represents a less densely populated secondary market.

Five regions were selected which differ in terms of total heating and cooling loads, and the balance between them. The populations of these regions are shown in millions in Figure 7.1 and Figure 7.2 for the years 1969 and 2000, respectively. The population distribution for the year 2000 is similar to the year 1969 except that the densities for all regions are increased. It should be noted that the greatest growth is expected in the Southeast and West.

For each region, a city for which climatic data is available was selected as representative of that region. The regions and representative cities are:

Region	Representative City
Gulf Coast	Mobile, Alabama
Southeast	Atlanta, Georgia
Northeast	Wilmington, Delaware
Great Lakes	Madison, Wisconsin
West	Santa Maria, California

FIGURE 7.1: 1969 DEMOGRAPHIC ANALYSIS FOR POCE SITES

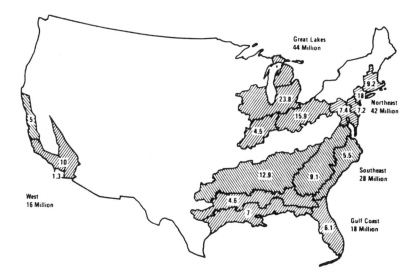

FIGURE 7.2: 2000 DEMOGRAPHIC ANALYSIS FOR POCE SITES

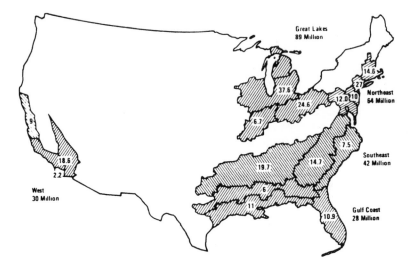

Table 7.2 indicates the population and the seasonal heating and cooling degree-days for the above regions and representative cities.

TABLE 7.2: REGIONAL DATA

Regional City	Population, Millions 1969/2000 yr	Seasonal Heating Degree-Days	Seasonal Cooling Degree-Days
Gulf Coast Mobile, Alabama	18/28	2000-1000 1612	3500-2700 3000
Southeast Atlanta, Georgia	28/42	4000-2000 2826	2500-1800 2250
Northeast Wilmington, Delaware	42/64	7000-4000 4910	2000-1000 1500
Great Lakes Madison, Wisconsin	44/69	8000-5000 7417	2000-1000 1250
West Santa Maria, Calif.	16/30	4000-2000 2934	1500-500 500

For single-family residence requirements, the regions are characterized by:

Gulf Coast — Heating is modest and cooling requirements dominate.

Southeast — Heating and cooling requirements are equal.

Northeast — Total heating and cooling requirements are high, with heating being about three times greater than cooling.

Great Lakes — Heating requirement is about six times that of cooling.

West — Heating requirement is about the same as that for the Southeast, but the cooling need is very low.

For most other buildings, the cooling load tends to dominate.

Buildings

The U.S. Department of Commerce (USDC) *Construction Review* groups buildings into 12 classes and provides statistics on the number of annual starts for each. Analysis of the data showed that a high percentage of all buildings would be encompassed if considerations were limited to the 6 classes in Table 7.3.

TABLE 7.3: PERCENTAGE OF BUILDING STARTS BY CLASS

USDC Class	Type	Percent
	Residential	
001	One Family Homes[1]	37
004	Five or More Family Buildings[2] (Low Rise Apartments)	37
300	Mobile Homes	21
	Total of Residential Construction	95
	Commercial	
015	Offices, Banks, Professional Buildings[3]	30
017	Schools/Educational Buildings[4]	4
018	Stores/Mercantile Buildings[5]	30
	Total of Commercial Construction	64

Notes:

1. Includes semidetached, row, and townhouses or townhouse apartments.

2. Each building contains five or more housing units having a common basement, heating plant, stairs, water supply and sewage disposal facilities, or entrance.

3. Includes post offices, court houses, etc.

4. Includes schools, colleges, libraries, gymnasiums, etc.

5. Includes stores, restaurants, laundry and dry cleaning shops, animal hospitals, etc.

There are 1 to 2 million building starts annually, and each building differs from the others in some way. To make the analysis tractable, a typical design and construction for each of the above building types in each of the five selected regions was determined. The characteristics included usage, total floor area, wall and roof dimensions, windows, ventilation, and insulation. It was recognized that future insulation practice for residential and commercial buildings would probably be improved over existing and past practices. Therefore, the heat gain/loss computations accounted for the anticipated insulation improvement.

The required thermal loads for space heating and cooling and hot water heating were calculated on an hour-by-hour basis for the 8,760-hour year, utilizing detailed weather data for each of the representative cities. The calculated annual and peak hourly heating and cooling load for each of the typical buildings in the five regions is presented in Table 7.4.

TABLE 7.4: BUILDING TYPES, PLAN AREAS, AND LOADS

	Building Plan Area Sq.ft.	Annual Heat Load 10^6 Btu	Annual Cooling Load 10^6 Btu	Annual Water Heating 10^6 Btu	Peak Hrly Htg. 10^3 Btu	Peak Hrly Cool 10^3 Btu
Atlanta,Ga. 1956						
Office	10,000	282	1912	91	815	1578
S.F. Home	1,560	36	45	26	35	36
Mobile Home	845	22	35	20	23	25
M.F. Apt.	7,424	129	508	205	230	301
School	130,000	1928	2978	1810	3942	5012
Store	1,400	26	101	1	107	95
Mobile, Ala.1963						
Office	10,000	176	1489	91	647	1197
S.F. Home	1,600	10	60	26	29	33
Mobile Home	845	11	47	20	24	24
M.F. Apt.	7,488	68	690	205	272	372
School	130,000	1012	4414	1810	3824	5607
Store	1,400	13	131	1	110	119
Santa Maria,Cal.1957						
Office	10,000	144	2423	91	938	1751
S.F. Home	1,632	21	14	26	19	25
Mobile Home	845	15	13	20	14	18
M.F.Apt.	7,680	114	302	205	145	279
School	130,000	1303	1237	1810	2860	2559
Store	1,400	9	87	1	41	55
Wilmington,Del.1959						
Office	10,000	479	1463	91	1070	1680
S.F. Home	1,144	58	41	26	54	32
Mobile Home	845	34	25	20	29	24
M.F. Apt.	7,680	209	385	307	340	288
School	130,000	3032	2039	1810	4356	5905
Store	1,400	31	82	1	145	104
Madison, Wis.1956						
Office	10,000	775	1234	91	1344	1783
S.F. Home	1,612	108	26	26	71	29
Mobile Home	845	48	21	20	30	26
M.F. Apt.	7,200	197	306	205	262	284
School	130,000	4517	1589	1810	5509	4869
Store	1,400	47	70	1	227	121

SOLAR SYSTEMS

Operational Requirements

It is economically impractical to design solar systems to provide 100% of the heating and cooling requirements of buildings by solar energy alone. The required size of solar collector to absorb, and storage subsystem to hold, enough energy to carry the building through protracted periods of cloudy days would be so great that the costs would be prohibitive. Solar systems will, therefore, incorporate conventional components such as gas- or oil-burning furnaces as auxiliary systems. The requirements and costs for solar systems which would provide 50 or 80% of the total annual energy needs were analyzed in this program, on the basis that a conventional auxiliary system would provide the balance.

Solar systems are more sophisticated that the conventional heating, ventilating, and air conditioning (HVAC) systems; and strong attention to system reliability and maintenance is essential. To provide a baseline and competitive goal for operation of solar systems, reliability data for conventional subsystems was obtained. The details are summarized in Table 7.5. It should be noted that a gas furnace has the best reliability, but that gas-fired absorption air conditioners have the poorest.

Generally residential heating and air conditioning systems have a useful life of about 15 years. Thus the solar and conventional systems in this study have been analyzed on a 15-year-life-cycle basis.

TABLE 7.5: RELIABILITY DATA FOR RESIDENTIAL HEATING AND COOLING SYSTEM COMPONENTS

Type	Yearly Service Call Rate Calls/Unit/Year
Gas Furnace	.42
Oil Furnace	1.16 - 1.80
Gas Air-Conditioning (Ammonia Absorption)	2.91 1.89*
Electric Air-Conditioning (Vapor Compression)	1.08
Heat Pump	1.38

*Based on a smaller sample for first-year operation

A thorough review was made of existing literature on solar systems, components, and techniques for heating and cooling of buildings. Those systems that could satisfy the requirements for a Proof-Of-Concept-Experiment and could be manufactured in volume quantities to meet the needs for the developing market were selected for detailed analysis and evaluation.

Conceptual diagrams of these systems, solar heating only, solar-assisted heat pump, and solar heating and cooling (absorption), are shown in Figures 7.3, 7.4, and 7.5. The performance of each of these systems was analyzed to determine the percentage of auxiliary energy required as a function of collector size.

FIGURE 7.3: SOLAR-POWERED HEATING ONLY

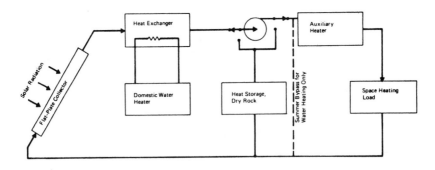

FIGURE 7.4: SOLAR-ASSISTED HEAT PUMP

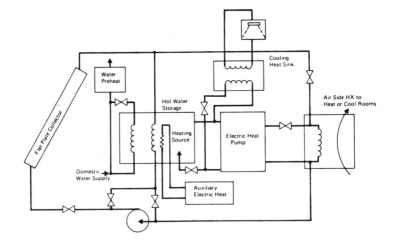

FIGURE 7.5: SOLAR HEATING AND COOLING WITH ABSORPTION AIR CONDITIONER

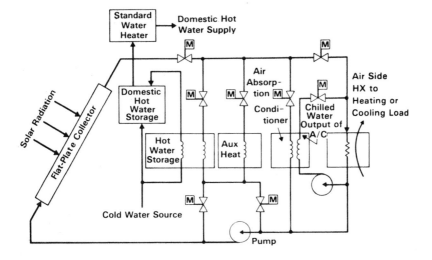

Region-Building-Solar System Combinations

The three selected solar systems were analyzed for each of the six building types and five regions, as tabulated in Table 7.6. Thus, a total of ninety region/building/solar system combinations have been analyzed and evaluated in this program.

TABLE 7.6: REGION-BUILDING-SOLAR SYSTEM COMBINATIONS

Cities	Buildings	Selected Solar Systems
Wilmington, Delaware	Single family residences (1,600 ft²)	Solar heat only
Madison, Wisconsin	Multifamily low-rise apartments	Solar-assisted heat pump
Atlanta, Georgia	(five or more families) (14,400 ft²)	Solar heat & cool
Mobile, Alabama	Mobile homes (832 ft²)	(absorption)
Santa Maria, California	Offices (35,000 ft²)	
	Schools/educational buildings	
	(210,000 ft²)	
	Stores (1,400 ft²)	

Collectors

The unique and essential component of all solar systems is the collector. Focusing collectors do not absorb any more energy than flat plate, nonfocusing configurations but, by concentrating the energy into a smaller area, can achieve temperatures as high as several hundred degrees Fahrenheit. In contrast, flat plate collectors absorb and utilize both the direct and diffuse components of solar radiation whereas focusing collectors function only with direct radiation. For heating-only requirements, flat plate collectors using liquid as the working fluid can generate temperatures of about $150°$F without difficulty and with good efficiencies.

Careful system design can also enable flat plate collectors to generate the temperatures of $180°$ to $200°$F required to drive absorption cooling systems although at reduced collector efficiencies. Flat plate, nonfocusing collectors were chosen from considerations of performance requirements, costs, and simplicity.

The collector types evaluated included those using liquid and air as working fluids. Designs involving one to four glass covers and both selective and nonselective absorbing surfaces were considered. Heat losses from the collector were determined for typical winter operation ($30°$F ambient temperature and $140°$F collector temperature) and typical summer operations ($90°$F ambient and $200°$F collector temperature). A level of 317 Btu/hr ft^2 insolation was used in determining collector efficiencies.

In view of uncertainties in the production processes and in the long-term stability of selective surfaces, a collector based on a nonselective surface with two glass covers was chosen for the heating-only system. Figure 7.6 illustrates the detail of such a design. When higher temperature operation for cooling is required, the use of a selective surface or a more complex physical design may be economically justified.

Determining the collector size necessary to provide the desired amount of energy must be carefully computed. Whereas the sizes of conventional heating and cooling equipment systems are calculated for maximum peak hourly demand, a solar system is calculated on an annual basis. The latter calculations involve annual solar insolation data for the locality, as well as the heating and cooling load of the specific building. An error in determining the proper collector size would not only cause poor performance but would be very costly. An error of 100% in sizing a conventional furnace results in a cost differential of only $175, but a similar error in sizing of the collectors (assuming a collector price of only $4 per square foot) costs $2,000.

Utilizing insolation and weather tape data for the selected five regions and the structural and use characteristics of the six classes of buildings, the collector size required for various levels of solar dependence or auxiliary fuel use was determined with the assistance of a computer. Table 7.7 summarizes the results for 50 and 80% solar dependency (50 and 20% auxiliary fuel). It should

FIGURE 7.6: MODULAR SOLAR COLLECTOR

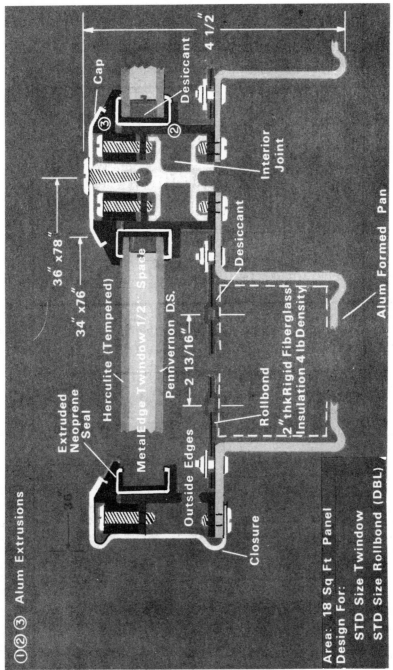

① ② ③ Alum Extrusions

Cap

Desiccant

4 1/2"

Interior Joint

Desiccant

Alum Formed Pan

36" x78"

34" x76"

2 13/16"

Extruded Neoprene Seal

Herculite (Tempered)

MetalEdge Twindow 1/2" Space

Pennvernon D.S.

Rollbond

2" thk Rigid Fiberglass Insulation 4 lb Density

Outside Edges

Closure

Area: 18 Sq Ft Panel
Design For:

STD Size Twindow

STD Size Rollbond (DBL)

be noted that, to provide 50% solar heating only for a single-family residence in the West (Santa Maria), only 160 square feet of collector is required, whereas 1,200 square feet is necessary in the Great Lakes (Madison, Wisconsin) region.

TABLE 7.7: COLLECTOR AREA REQUIRED FOR SINGLE-FAMILY RESIDENCES

Region	50% Solar Dependency			80% Solar Dependency		
	HO	HP	HC	HO	HP	HC
Northeast (Wilmington)	440	400	680	1,120	1,120	1,380
Southeast (Atlanta)	240	160	560	820	(2,000)	1,260
Gulf Coast (Mobile)	120	100*	580	240	(---)	1,200
Great Lakes (Madison)	1,200	480	860	(---)	(---)	(---)
West (Santa Maria)	160	100*	100*	220	120	160

Key: Solar collector area (ft^2) required for:

HO	-	Heating Only
HP	-	Solar-Assisted Heat Pump
HC	-	Heating and Cooling (Absorption)
*	-	Required for domestic hot water only
( )	-	Collector area exceeds available roof area

Storage Subsystems

Techniques which utilize latent heat of fusion (such as in eutectic salts) continue to show promise of reducing the volume required for energy storage. However, none of these techniques has proven dependable or consistent in performance over many cycles of heating and cooling and extended periods of time. Sensible heat storage is the most reliable technique for near-term application.

Dry rock is considered the most practical heat storage medium for air-heating collectors, and water is the choice for liquid systems. These were selected on the basis of simplicity, availability, reliability, and maintainability, as well as economics.

Cooling

Utilizing solar energy to perform the cooling function is much more difficult than for heating. Six systems were evaluated: absorption systems, adsorption systems, jet ejector systems, Rankine cycle–inverse Brayton cycle, Rankine cycle–vapor compressions, and night radiation schemes.

These evaluations disclosed that more extensive operating and manufacturing experience has been obtained with lithium bromide-water and ammonia-water absorption air conditioners having capacities over the complete range of interest than with any other complete system alternative. Adsorption systems involving drying and subsequent evaporative cooling will require extensive development before they can be applied with confidence.

Night radiation schemes were found to be dependent on local humidity and ambient temperature conditions, and their application is geographically constrained to an undesirable extent.

Solar-Assisted Heat Pump

A heat pump is basically a modified air conditioning system that has the flexibility to interchange functions between the evaporator and the condenser, allowing it to either heat or cool the desired space. Figure 7.7 represents an example of a heat pump in the heating mode. During the summer, it operates as a conventional cooling system. Cool air is delivered to the conditioned space, and the heated air from the condenser is exhausted outdoors.

In the heating mode, during the winter, the evaporator cools the outdoor air by absorbing heat at the low ambient temperatures. The compressor then pumps the heat to a higher temperature which is delivered to the space to be warmed. The heat delivered is several times the electric resistance heat equivalent of the input power.

The coefficient of performance (COP) of the heat pump varies with operating conditions. The variable with the greatest effect on the COP is the temperature of the outside air (or the water) to which heat is rejected in the summer, and from which heat is drawn in the winter. During the winter, solar energy can be employed to maintain higher input temperatures to the heat pump, resulting in a higher coefficient of performance. During the summer, solar energy is not utilized, and the system functions as a conventional air conditioner.

ECONOMIC CONSIDERATIONS

Preliminary Cost Study

The life-cycle costs of candidate solar systems were estimated and compared among themselves and with conventional systems to identify those having the

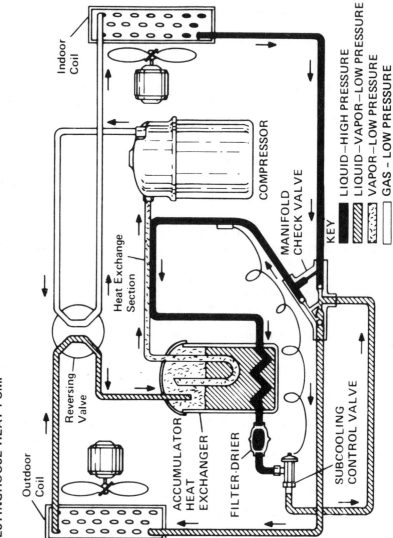

FIGURE 7.7: WESTINGHOUSE HEAT PUMP

greatest potential on a cost-effective basis. To determine costs, the three candidate systems were designed in sufficient detail to identify and define all of the hardware, subsystems, transportation, labor, and markup to permit an estimate of the installed system cost. Concurrently, investment costs of comparable conventional systems were established, and operating and maintenance costs were determined for all systems.

Since solar systems are not designed for 100% solar dependency, they require supplementary conventional equipment and fuels to ensure 100% performance during extremely cold or hot periods and during prolonged cloudy weather. The costs of the supplemental equipment and fossil fuel, as well as the fuel cost for conventional systems, were determined.

Using the investment, operating and maintenance, and fuel costs, the 15-year life-cycle costs were calculated for solar and conventional systems. The life-cycle cost was determined for solar systems designed to provide one-half of the heating and cooling load from solar energy and one-half from conventional fuel (i.e., 50% solar dependency). The life-cycle costs were also calculated for systems designed for 80% solar dependency.

In addition, calculations were made for each of two dates, 1975 and 1985, using constant 1973 dollars. Future costs were discounted 8%, but included an annual inflation factor of 5% on equipment and maintenance costs, and a fuel escalation cost of 7% per year. Thus, the life-cycle cost for each system, each degree of solar dependency, and each date is equal to the investment cost plus the sum of the discounted annual charges.

Examples of these cost estimates are shown in Table 7.8. This table contains the estimated investment and life-cycle costs of each of the three 50% solar-dependent heating and cooling systems and two comparable conventional systems, in a single-family residence, in each of two regions for both 1975 and 1985.

Life-cycle costs are useful in comparing dissimilar equipment performing similar tasks. Systems having the lowest life-cycle cost are the most economic. For example, in Table 7.8 it is seen that, in 1975, for a single-family residence in the northeast region, the life-cycle cost of a solar-assisted heat pump providing 50% solar dependency is $2,330 more expensive than a conventional HVAC. But, by 1985, the solar-assisted heat pump is within $200 of the conventional HVAC.

Thus, given an annual expectation of 5% wage inflation, a 7% growth in the cost of energy, and a 2.5% reduction in the cost of solar systems (characteristic of many new industries), solar heating and cooling systems should become generally competitive in many regions of the United States by 1985. Analysis of the life-cycle costs are calculated in this study discloses that very few solar system cases compare favorably with conventional systems in 1975. But, again by 1985, many solar system cases are more cost-effective than conventional systems.

TABLE 7.8: SINGLE-FAMILY RESIDENCE, INVESTMENT AND LIFE-CYCLE COST ($) 50% SOLAR DEPENDENCY

Heating/Cooling Region and System	1975		1985	
	Investment	15-Year Life-Cycle	Investment	15-Year Life-Cycle
West Coast				
Santa Maria, Calif.				
Solar Heating and Cooling	3650	5820	2840	6640
Solar Heat Pump	-	-	-	-
Solar Heating Only	2540	3500	1970	3730
Conventional Heating and Cooling	2220	5020	2850	8100
Conventional Heating Only	1110	2530	1420	4110
Northeast				
Wilmington, Del.				
Solar Heating and Cooling	8810	12700	6850	13800
Solar Heat Pump	4800	9930	3740	13400
Solar Heating Only	4220	5860	3290	6300
Conventional Heating and Cooling	2220	7600	2850	13200
Conventional Heating Only	1140	3420	1460	5870

Capture Potential

The amount of conventional fuel that can be conserved through substitution of solar energy is directly dependent on the number and energy demand of buildings of various classes that would adopt solar systems. The fundamental question is, how much of the heating and cooling market can solar systems capture?

The willingness of a home buyer or commercial building developer to choose solar over conventional systems is greatly influenced by economic considerations. In view of the increased initial investment required for solar systems over conventional systems, a decision to install a solar heating and cooling system would be motivated by the expectation of reducing future operating costs by an amount at least equal to that of the increased investment. The maximum amount a rational consumer might be willing to invest in a solar system should be repaid by the reduction in fuel bills he would experience in combination with any tax incentives. Other factors may influence the selection of solar systems but, in the final analysis, economics should dominate the majority of decisions on a long-term basis.

In determining the competitive equivalent of solar versus conventional systems, the principal factors involved for each building in the selected geographical regions include: heating, hot water, and cooling requirement (Btu/yr); annual solar insolation (Btu/yr); solar collector size (ft^2); solar collector installation costs ($/ft^2$); competitive fuel (gas/oil/electricity) prices for specific time periods ($/million Btu); and cost of conventional system ($).

Of the above, the solar collector installed costs versus the price of competitive fuels dominate the equation of solar energy economic feasibility. These prices are time dependent. It is expected that, in the next several years, solar collector costs will decrease as a result of design and manufacturing improvements and that electricity, gas, and oil prices will increase.

The downward push of solar collector costs and upward pull of conventional energy prices will result in solar systems reaching a competitive position in respect to conventional systems. This trend may be accelerated by the application of tax incentives and other government stimuli to improve the economic advantages of solar systems.

Impressive progress has been made recently in the design improvement and cost reduction of the several solar collectors that are being manufactured. The collector manufactured by PPG Industries is particularly attractive. It uses two tempered glass covers and Roll-Bond as the absorber plate. For new construction, collectors can become a part of the structural walls or roof. The credit in labor and materials for that portion of the buildings replaced by solar collectors could substantially offset the collector installation costs in most cases. Thus, the installed collector cost is essentially equivalent to the cost of the collector itself.

The year in which economic feasibility will be attained will vary, depending on the region of the country, the level of solar dependency, and the type of solar systems, as well as all the factors listed above. Figure 7.8 is an example for a solar heating-only system providing 50% of the heat from solar energy and 50% from gas or oil. For this chart, as a simplification, current and projected national fuel price averages are indicated on the ordinate. Figure 7.9 is a similar chart for heating and cooling.

The curves, a simplification of a more complex set of conditions, represent the crossover from noneconomic to economic feasibility for the indicated systems. Points below the curves are conditions of nonfeasibility and those above represent conditions of economic feasibility. It can be expected that collector costs will be reduced as a consequence of product improvement and production

FIGURE 7.8: ECONOMIC FEASIBILITY OF SOLAR SYSTEMS—HEATING ONLY

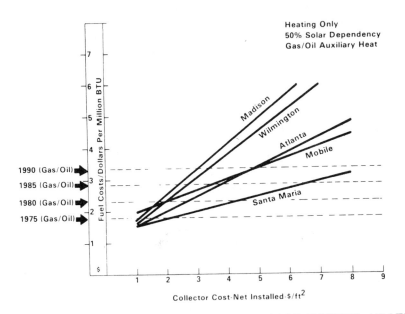

FIGURE 7.9: ECONOMIC FEASIBILITY OF SOLAR SYSTEMS—HEATING AND COOLING

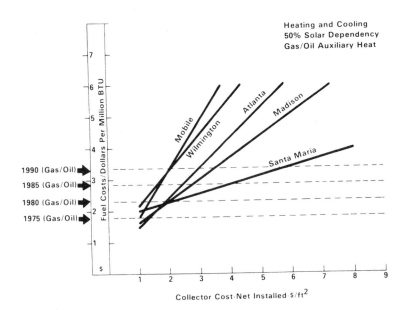

rate. The following are considered reasonable projections for the indicated time periods:

Period	Collector Cost/Square Foot
1975-1980	$5.50
1980-1985	$3.00
1985-2000	$2.00

On the basis of these anticipated collector costs, the time periods during which the selected solar systems reach economic feasibility in the various regions have been computed and the results indicated in Table 7.9. For the purpose of this table, regional energy prices rather than national averages were utilized.

The annual total capture potential for buildings, solar equipment sales, and amount of fuel saved in 1980, 1985, and 1990 for the combined regions was computed and is summarized in Table 7.10.

TABLE 7.9: TIME OF REACHING ECONOMIC FEASIBILITY—SINGLE-FAMILY RESIDENCES

System	Region	Competing Fuel/System		
		Fossil	Elec/Res	Elec/HP
Solar Heat	Madison	1985–90	1985–90	2000
	Wilmington	1985–90	1980–85	1990
	Atlanta	1985–90	1980–85	1980–85
	Mobile	1980–85	1980–85	1975–80
	Santa Maria	1975–80	1975–80	1975–80
Solar Assisted Heat Pump	Madison	1985–90		2000
	Wilmington	1985–90		2000
	Atlanta	1985–90	(1)	1990
	Mobile	1985–90		1995
	Santa Maria	1980–85		1985–90
Solar Heat and Cool	Madison	1985–90	1980–85	2010
	Wilmington	1990	1995	2020
	Atlanta	1990	1990	2015
	Mobile	1995	2015	2020
	Santa Maria	1980–85	1980–85	1985–90

Note: (1) Not applicable

(2) Collector costs are: 1975 – $5.50/ft^2
1980 – $3.00/ft^2
1985 – $2.00/ft^2

(3) Energy costs are those projected for the region and year indicated

TABLE 7.10: SUMMARY OF CAPTURE POTENTIAL

	1980	1985	1990
Buildings			
Single-Family Residences (units)	16,000	129,000	174,000
Multifamily Residences (dwelling units)	227,000	344,000	308,000
Nonresidential (square feet)	98,000,000	278,000,000	410,000,000
Solar Equipment Sales (In addition to conventional HVAC Components)			
Single-Family Residences	$22,000,000	$226,000,000	$357,000,000
Multifamily Residences	$67,000,000	$205,000,000	$238,000,000
Nonresidential	$79,000,000	$339,000,000	$580,000,000
Fuel Saved	1,040,000	4,820,000	6,990,000
(Annual increment-bbl oil/year)			

The amount of conventional energy, expressed as equivalent barrels of oil, that could be saved by the use of solar energy deployed as a function of time is such that the near-term detail indicates a slow rise in energy economy reaching 50 million barrels of oil per year by 1990. The initial slow increase is in part a consequence of the normal limited construction rate of new buildings compared to the large inventory and long life of existing buildings. Saturation in terms of energy saved per year would occur in about 50 years. When that point is reached, the amount saved will be very large. In fact, it will be greater than the present automobile consumption in equivalent barrels of oil, and the dollar market will approach ten billion 1973 dollars.

SOCIAL, ENVIRONMENTAL AND INSTITUTIONAL CONSIDERATIONS

Social and Environmental

An assessment of the social and environmental acceptability of solar energy systems was conducted in specific problem areas, including:

The social acceptability at the user level of solar heating and cooling of buildings, including the effects of cost, appearance, and financing.

The environmental acceptability of the use of solar energy

for heating and cooling of buildings, including pollution, noise, and aesthetics.

The impact of solar heating and cooling of buildings upon heating, ventilating, and air conditioning manufacturers and upon the labor force associated with them.

The wide-scale application of solar heating and cooling of buildings through a building systems approach, using both architectural and engineering inputs.

The effort was divided into two major parts: delineation of constraints and acceptability of systems.

The survey technique was employed to assemble basic data. Seven relevant population groups were surveyed: architects, builders, labor groups, manufacturers, energy suppliers, financiers, and potential consumers. After the data produced by the surveys were analyzed, in-depth "one-on-one" interviews were utilized to assess the acceptability of solar heating and cooling systems. The findings of these efforts are generalized in the following paragraphs.

The respondents favor the idea of solar heating and cooling, but their enthusiasm drops off sharply as the system becomes more expensive.

Architects favor emphasizing all types of buildings in introducing solar heating and cooling. Schools, commercial buildings, and office or professional buildings were singled out, however, as requiring special emphasis. Architects clearly prefer fitting systems to newly designed buildings as opposed to retrofitting existing buildings.

Reactions of builders to the proposed solar heating and cooling system range from skepticism to qualified acceptance. There is general concern that, without extensive merchandizing backed up by documented success of solar systems, buyers would tend to stick with traditional systems. Costs, aesthetics of the solar collector, and location of the storage tanks are major concerns.

Labor groups are in general agreement with other populations. Increased construction time required for installation of a solar system is marginally acceptable, and they perceive the need for a modest amount of special training.

The manufacturers interviewed are generally quite favorable. Several said that their firms are peripherally engaged in assessments of the feasibility of solar energy usage.

Of the seven groups canvassed, energy suppliers are the least enthusiastic about the proposed solar model. Their objections centered on high cost, technological complexity, constantly fluctuating demand on their services, and general skepticism with respect to consumer acceptance.

Financiers believe that a solar supplement would have no direct adverse effect on financing and could possibly improve it. It is felt that solar heating and cooling systems will enhance the salability of a building, and that solar unit sales should experience a strong growth rate of between 10 to 50% per year once these systems are in use.

Consumers indicate a decreasing acceptability with increasing costs and decreasing fuel cost reductions, as would be expected. The potential consumer finds an additional cost (over conventional system cost) for solar systems of $1,000 to $2,500 very acceptable in all cases. An additional cost of $2,500 and $5,000 would be on the border line between unacceptability and acceptability. Any additional costs over $5,000, regardless of fuel cost reduction, are rated very unacceptable.

High costs for solar heating and cooling systems were found to be more acceptable in new buildings. The cost of solar heating for large buildings such as schools should not exceed 15% of the total building cost. A solar collector site on the roof of the building is judged to be acceptable by almost all populations. An off-site center location also receives quite acceptable ratings, especially from consumers.

Utilization Planning Considerations

The effective utilization of the results of the NSF Solar Heating and Cooling of Buildings Program is dependent on a comprehensive, active plan for that purpose.

Architects and Engineers: The utilization plan would include a program to develop, refine, and disseminate standardized design procedures and data, working through the architects and engineers involved with the POCE designs and through organizations such as the American Institute of Architects (AIA) Research Foundation, Inc. and the American Society of Heating, Refrigeration, and Air Conditioning Engineers, Inc. (ASHRAE).

As the key technical resources utilized in the planning and design of buildings, architects and engineers represent a fundamental resource in institutionalizing both the technical capacity to incorporate solar systems into building designs, and a knowledgeable source that can promulgate utilization through recommendations to owners and investors.

Recognizing the important interaction between land-use planning and architectural design, a parallel program to address the development of model land-use planning guidelines should be conducted with organizations such as the American Institute of Planners (AIP) and the American Society of Planning Officials (ASPO).

Building Contractors: Many building contractors have the capacity, but inadequate information, to respond to solar construction and installation requirements in nonresidential buildings. An information program to assist these contractors and subcontractors to adapt their procedures to these requirements should be initiated through national contractor organizations, such as the Associated General Contractors (AGC).

A program of builder education and involvement in specifying design refinement objectives would be initiated with the assistance of the residential POCE builders and the NAHB Research Foundation, Inc.

Code and Safety Requirements: Existing equipment, building, and safety codes do not appear to require changes for the introduction of solar systems. However, the experience in the construction industry amply demonstrates that the absence of the negative prohibition does not necessarily equate to the presence of positive interpretations by the various local code bodies involved in the review and approval of construction projects.

The utilization plan includes an intensive effort to ensure that designers are provided adequate information to avoid creating problems, and that code officials are provided adequate information from qualified and reliable sources to enable them to interpret codes applicable to solar system designs. This program would be conducted in conjunction with the various national model codes groups, professional and trade standards groups, and insurance and safety standards groups.

Channels of Distribution: Analysis of the manufacturing, distribution, and installation requirements of solar systems indicates that the existing channels of distribution of heating and cooling equipment offer clear advantages over any other alternative for at least an initial period. The preliminary plan is based upon making the investments necessary to utilize these channels in ways that minimize the need for external investment. At the same time, it would maximize the flexibility for achieving the evolution from early generations of equipment to more completely packaged systems in the longer range.

Recognizing the need for more incentive in adaptation by the residential channels of distribution, a proportionately larger investment in providing external resources and packaging designs will be required.

Customer Service: The importance to buyers of adequate service and parts availability will be a significant factor in decisions to utilize solar systems. Customer servicing requirement plans follow closely the distribution channel plans. The plan gives recognition to the need for in-place servicing capabilities and replacement parts at the time initial systems are deployed. It also recognizes the probability that the need for servicing will be greater on initial systems, and that the costs of establishing and maintaining this state of readiness will be greater than for conventional systems.

The objective of the plan would be to ensure that both in-warranty and out-of-warranty service would be locally available at costs competitive with conventional systems from sources that normally provide similar kinds of service to conventional systems.

Government Policies and Incentives: A major element of the utilization plan is the establishment of a coordinated set of government policies and incentives that would create an initial market and foster the subsequent private investment to establish a viable industry. Since widespread economic competitiveness will depend upon both significant reductions in installed collector costs and increases in conventional fuel costs, external investment will be required if early and widespread utilization is to be utilized.

Buyers and Homeowners: Before widespread utilization of solar energy can occur, there must be a broad base of buyer understanding of the consumer benefits of installing solar heating and cooling systems. The proposed program can create that understanding and become the basis for broad public acceptance. The plan is based on working with key buyer-oriented trade and professional groups in nonresidential construction, and key consumer and environmental groups in residential construction.

By working with these groups, starting with the POCE and continuing during precommercialization, realistic assessments of readiness and benefits can be established and disseminated. Simultaneously, the spin-off information flowing from the architect/engineering professionals, the builders, and the early users involved with POCE can provide supporting evidence to potential buyers.

RECOMMENDATIONS

A Research Project

It is proposed to design, procure, install, monitor, and analyze the performance of a large-scale solar heating and cooling system of the following characteristics: solar collector area of approximately 10,000 square feet, 100-ton absorption cooling capacity, commensurate space heating capability, and hot water heating for domestic use.

Research Objectives: The technical objectives of the experiment are:

> To reveal and resolve unanticipated system problems associated with large-scale heating and cooling systems that cannot be completely foreseen in theoretical studies of the design, fabrication, installation, and operation of a large-scale solar heating and cooling system.

> To demonstrate the performance of a well-engineered, low-cost, assembly-line-produced, commercial collector of high performance.

> To operate the above collector at the high fluid temperature (200°F) necessary for absorption cooling and to store the energy at these temperatures. A collector and storage system designed for opera-

tion at such temperatures has not previously been demonstrated.

To optimize the collector tilt angle for seasonal heating and cooling at a specific site in the Southeast.

To demonstrate solar heating and cooling of an entire building as distinct from a small section.

Research Plan: The city of Atlanta, Georgia, typifies the climatic and insolation characteristics of the southeastern region of the United States, wherein the seasonal heating and cooling needs are very nearly balanced, and the population will double by the year 2000. An elementary school in Atlanta is proposed as the site of this demonstration project.

The Atlanta schools are active throughout the year on a four-semester schedule, and are utilized at night and on weekends for community activities. The use of an Atlanta elementary school would result in a high level of experiment utilization and exposure, and the market potential in this area is large. A solar system including 10,000 ft^2 collector area, water thermal storage of 24,000 gal, and a 100-ton LiBr absorption chiller was recommended to interface with the forced-air or circulated-hot-water heating system. The resulting system should be designed to operate automatically to satisfy the total heating and cooling needs of the building.

System Configuration: The system configuration recommended for this project is that shown in Figure 7.5. Since the system is to be located where the temperature seldom drops below the freezing point, a design option not requiring the use of antifreeze in the collector loop should be considered. On the rare occasions when the temperature does approach freezing, hot water from storage can be circulated through the collector at a very low rate to prevent freezing.

This can be accomplished by an automatic sensing and control system and a small auxiliary pump. The elimination of a heat exchanger between the collector and hot water storage will permit slightly higher storage temperatures, which are desirable for powering the absorption air conditioner, and will result in a cost saving.

Collector Design: A collector suitable for adaption to this project is manufactured by PPG Industries, Inc. The collector module consists of a Roll-Bond non-selective collector plate and two layers of tempered glass, mounted in a stainless steel frame. Each collector assembly measures approximately 32 x 74 inches, with 17 square feet of collector area. About 588 of these panels will be required to provide a total collector area of about 10,000 square feet. During a period of approximately three hours centered about noon on a clear day, this collector area will supply about 2.0×10^6 Btu/hr, which is slightly more than the peak thermal energy requirement of a 100-ton absorption air conditioner.

Incident to the design phase of this experiment, the advantages of orienting the collectors flat on the roof (as opposed to a tilt orientation to maximize insolation) should be thoroughly explored.

Absorption Air Conditioner : An absorption air conditioner suitable for this project is the Arkla WF-1200 water-fired unit. It is the only commercially available unit designed to operate directly from hot water at a temperature between $200°$ and $245°F$. The unit is rated at 100 tons of refrigeration, with a coefficient of performance of 0.71.

Thermal Energy Storage: For a solar heating and cooling system, where the load is to be predominantly cooling, energy usage is closely correlated with insolation; therefore, relatively short-term (on the order of a few hours) storage is usually provided. It is recommended that 24,000 gallons of water storage be provided and additional heat made up, when required, from the building's water heater.

Performance Analysis: The performance of the solar heating and cooling system should be monitored and analyzed through at least one year of operation. From this analysis, design optimizations and design and performance extrapolations to other-size solar systems should be performed. System and component reliability and maintainability data should be analyzed to apply lessons learned to Proof-of-Concept-Experiments to follow.

Cost: The estimated cost for this project is $500,000.

Proof-of-Concept Experiments

To serve as a stimulant and have a significant impact, the POCE should be an order of magnitude greater than previous efforts. It should involve testing each of the selected solar systems for the six classes of buildings in all of the regions. Not only performance but potential maintenance and reliability problems must strongly influence considerations of the scale of the experiments.

With these considerations in mind, it was judged that an experiment involving three hundred single-family residences and twenty-five larger buildings would be of the proper magnitude. A budget requirement for this size experiment is estimated to be about $26 million.

The total number of combinations studied was 90 (three systems times six buildings times five regions). The POCE matrix recommended was based upon the following factors: present area population, year 2000 population, heating degree-days, cooling degree-days, assumed minimum viable experiment for single-family residences, and the assumption that non-single-family residential experiments should be approximately equal in dollar expenditure to single-family residential. The first four items above are shown in Table 7.11.

The incremental population times the sum of both heating and cooling degree-days is an indicator of the maximum potential heating and cooling load. Multiplying these by the average solar dependency (for heating only, heating and

TABLE 7.11: DEMOGRAPHIC AND THERMAL LOAD FACTORS RELATIVE TO POCE

Region (City)	1969 Population	2000 Population	Heating Degree Days	Cooling Degree Days[*]
Gulf Coast (Mobile)	18 Million	28 Million	1612	4280
Southeast (Atlanta)	28 Million	42 Million	2826	3210
North East (Wilmington)	42 Million	64 Million	4910	2140
Great Lakes (Madison)	44 Million	69 Million	7417	1790
West (Santa Maria)	16 Million	30 Million	3934	715

[*] Increased 43 percent over sensible degree-days to account for dehumidification requirements.

TABLE 7.12: RELATIVE MAXIMUM CAPTURE POTENTIAL

(1) Region (City)	(2) Population Increment	(3) Total Degree Days	(4) Ave. Solar Dependency	Product (2)x(3)x(4)	Relative Capture
Gulf Coast (Mobile)	10 Million	5892	.68	40,070	11.1%
Southeast (Atlanta)	14 Million	6036	.74	62,530	17.3%
Northeast (Wilmington)	22 Million	7050	.59	91,510	25.4%
Great Lakes (Madison)	25 Million	9207	.51	117,390	32.6%
West (Santa Maria)	14 Million	4649	.96	49,040	13.6%
				Total	100.0%

cooling, and heat pumping at a collector area equal to 50% of plan area), the resulting number will roughly indicate the relative maximum capture potential. The results of this calculation are shown in Table 7.12.

The size of each experiment in the respective cities (and their environs) was rounded to 10, 20, 30, 30 and 10%. The initial assumption was further made that a minimum viable experiment in any one city should be 30 residences to allow for system variation, residential design variation, and location variation

(rural versus urban). Furthermore, consideration was given to the character of
the loads (heating versus cooling). An additional judgement was made as to
the experimental mix for each city. For example, since the load in Mobile is
primarily cooling, no heating-only experiments are recommended there. The
preliminary single-family residential recommendations are shown in Table 7.13.

Also included in Table 7.13 are recommendations with regard to mobile homes.
Preliminary cost analyses have shown that, in all locations except Santa Maria,
the incremental system cost for mobile homes is too high to make such systems
acceptable since the purchase of mobile homes is strongly influenced by comparatively low initial cost. However, in Santa Maria the incremental cost appears
to be in the range of $1,000 to $1,500 and, therefore, ten heating-only experiments in mobile homes for the Santa Maria region are recommended.

TABLE 7.13: SINGLE-FAMILY RESIDENTIAL POCE

| Region (City) | Mobile Homes (HO) | Other Single-Family Residences | | | |
		Heating Only	Heating and Cooling	Heat Pump	Total
Gulf Coast (Mobile)	0	0	15	15	30
Southeast (Atlanta)	0	10	25	25	60
Northeast (Wilmington)	0	35	35	20	90
Great Lakes (Madison)	0	25	40	25	90
West (Santa Maria)	10	10	15	5	30
				Total	300

With regard to nonresidential structures, it was decided that there should be
at least one POCE experiment for each building type in each city. For schools,
two in each city have been recommended because they represent excellent candidates for retrofitting and because of their public ownership. The capture potential results stressed the importance of air conditioning for large buildings. Therefore, the POCE recommendations for large buildings include only heat pumps
and heating and cooling experiments, with a 1.5:1 ratio between the two as
being representative of the estimated degree of technical feasibility.

Additional factors employed in determining the mix of building types were
economic feasibility, capture potential, effectiveness, reliability, availability,
maintainability, manufacturability, and acceptability. The final tabulation is
shown in Table 7.14.

TABLE 7.14: RECOMMENDED PROOF-OF-CONCEPT-EXPERIMENT DISTRIBUTION

	Mobile Home	Single Family Residence	Multi-Family Residence	Office Building	Store	School
Atlanta, Georgia (Location 1)	0	20 HO 26 HP 20 HC 66	1 HP	1 HP	1 HC	2 HC
Mobile, Alabama (Location 2)	0	8 HO 18 HP 14 HC 40	1 HC	1 HC	1 HC	2 HC
Santa, Maria, California (Location 3)	10 HP	16 HO 24 HP 16 HC 56	1 HC	1 HP	1 HC	1 HP 1 HC
Wilmington, Delaware (Location 4)	0	16 HO 18 HP 14 HC 48	1 HP	1 HP	1 HP	1 HP 1 HC
Madison, Wisconsin (Location 5)	0	30 HO 30 HP 30 HC 90	1 HC	1 HP	1 HC	1 HP 1 HC
System Totals	10 HP	90 HO 116 HP 94 HC	2 HP 3 HC	4 HP 1 HC	1 HP 4 HC	3 HP 7 HC
Total Experiments	10	300	5	5	5	10

Note: HO = Heating only

HP = Solar-assisted heat pump

HC = Heating + absorption air-conditioning

Development of a Master Plan

The Proof-of-Concept-Experiments recommended would consist of 300 single-family residences and 25 larger buildings. Plans for preliminary system design, critical component design, fabrication, and test have been made. These plans will provide the base for detailed design, implementation, and operation of the POCEs as well as indicate critical subsystem research and development. Plans for system construction and testing have also been developed.

Critical Component/Subsystem Research and Development: A solar heating and cooling system will, in general, consist of a large number of components and subsystems which must operate in a very sophisticated integrated system under a control system which presents options of the utmost simplicity to the operator while, at the same time, providing comfort, performance, safety, and fail-safe protection of the system apparatus. These components and subsystems can be characterized as follows: available off-the-shelf, requiring straight-forward engineering design and manufacture, or requiring further research and development as to design and/or manufacture. Table 7.15 lists system components and subsystems in accordance with the foregoing classification.

TABLE 7.15: COMPONENT AND SUBSYSTEM CHARACTERIZATION

Off-the-Shelf	Straight-Forward Engineering & Design	Requiring Further Development
Air-Handling Equipment	Plumbing Systems	Control Systems
Cooling Towers	Heat Pumps	Collectors
Hydronic Units	Storage Systems	Absorption Refrigeration
Furnaces	Heat Exchangers	Integrated Packaged Units
Pumps		
Valves		

Absorption refrigeration in particular requires strong development effort. The absorption refrigeration system problems include: pump failures, system leaks (purge and refill), corrosion, hydrogen generation (purge and refill), valving failures, condenser fouling, and crystallization of LiBr.

In addition to these problems, which are sensitive to design and materials selection, problems peculiar to solar system operation will arise because of the wider variability of generator temperature available.

System Design, Construction, and Test: System design has been divided into three major categories: single-family residences—heat pumping and heating only, single-family residences—heating and cooling, and large buildings—heating and cooling.

Figure 7.10 is a master plan for the design, construction, and test of single-family residences. The plan is subdivided into three principle sections in which the individual elements overlap in scheduling. These are: Phase 1—system design, Phase 2A—POCE precursor, and Phase 2B—POCE.

Phase 1 would begin with a detailed determination of the design requirements at the component level and critical component development (some of which may have been undertaken as part of independent projects). The designs for solar heating only and solar-assisted heat pumps for the first group of single-family residences (Santa Maria) would be completed about 10 months after program start, and 5 months later for the last group (Madison). For solar heating and cooling, the designs would be completed in 17 and 20 months, respectively. On this basis, the construction of the first group of single-family residences would begin 18 months after start of Phase 1.

FIGURE 7.10: MASTER PLAN

Year of Program

| | 1 | 2 | 3 |

Design Requirements at Component Level

Critical Component Development
● Collectors
● Absorption Refrigeration
● Control (Non S.F. Residential)
● Integrated Packaged Units

S.F. Residential System Design (HO & HP)

S.F. Residential System Design (HC)

S.F. Residential Building Design

Heat Pump System Design

Heat Pump Development, Design & Proc.

Solar System Design & Procurement

Building Construction and System Installation

System Checkout

S.F. Residential HO (Detail Des.,Proc.,Install.)

S.F. Residential HP (Detail Des.,Proc.,Install.)

S.F. Residential HC (Detail Des.,Proc.,Install.)

Phase 1 - System Design

Phase 2A - POCE Precursor

Phase 2B - POCE (S.F. Residential)

Release Of System Design
▷ First Group
▷ Last Group

① First Group of Residences
② Last Group of Residences
△ Start Detailed Design
◇ HVAC Hardware Release
→ Building Start
✕ HVAC Equipment Delivery
● Occupancy

To obtain experience with single-family residences earlier than indicated in the above schedule, a POCE precursor, designated Phase 2A is proposed. The precursor program would consist of a total of 12 single-family residences (6 in each of two locations) using solar-assisted heat pumping for heating purposes and the inverted operation of the heat pump in the summer for cooling. It is believed that systems of this type could be constructed and made ready for occupancy within a 14-month period.

These systems would not meet all of the objectives of the POCE but would provide an early experience with special component development and medium-scale construction problems. The systems would not be optimized for cost benefit nor would they necessarily have the reliability nor maintainability of the later POCE's. The major development effort in the precursor program would be devoted to heat pumps suitable for use in solar systems.

Large Buildings: The POCE recommendations call for installation of 25 systems in large buildings including apartments, offices, stores, and schools. All of these would employ either a heat pump or an absorption air conditioner. An unspecified number would be retrofitted to existing buildings. Since each project will be custom designed, and separately and explicitly costed and scheduled, to satisfy the requirements of the specific project, it is not practicable to prepare preliminary schedules for them. It is envisioned that each project will entail a separate procurement.

The purpose of the Phase 1 effort (with regard to these larger custom-designed buildings) will be to provide modular component specifications and designs that will be acceptable to architects and engineers responsible for both the building design and its HVAC system.

Costs: The estimated costs for Phase 1 (Design), Phase 2A (Precursor), and Phase 2B (POCE) are summarized in Table 7.16.

TABLE 7.16: SUMMARY OF COSTS OF SOLAR EQUIPMENTS AND SYSTEMS

Phase 1

Single-Family Residence Design	$ 1,120,200
Large Building Design	1,087,500
Critical Component Development	1,632,000
Phase 1 Subtotal	$ 3,839,700

Phase 2A (Single-Family Precursor)

12 Single-Family Residential Systems	$ 903,200

(continued)

TABLE 7.16: (continued)

Phase 2B (POCE)

  Includes

300 Single-Family Residential Systems	
10 Mobile Home Systems	12,275,000
25 Large Buildings	
Monitoring, Data Collection, Etc.	5,540,000
Maintenance (5 years)	1,502,000
Contingencies	2,500,000
Phase 2B Subtotal	$21,800,000
Total	$26,542,900

TRW Study

The material in this chapter is from NSF-RA-N-74-022A, *Solar Heating and Cooling of Buildings (Phase 0)*, an executive summary prepared by TRW for the National Science Foundation in May 1974. Also contributing to this study were Arizona State University; DeLeuw, Cather and Co.; Arkla Industries; Foster Associates and Robert C. Lesser & Associates.

BACKGROUND AND APPROACH

The Phase 0 solar heating and cooling of buildings study was carried out in a logical and sequential manner. The methodology consisted of the following:

Eighteen climatic regions were identified (nine for the heating season and nine for the cooling season)
Five building types were selected
Fourteen cities were selected
Four system functions were defined
Building loads were calculated for hot water, space heating and cooling, and dehumidification
Three reference system designs were identified
System operation requirements were determined
Cost analyses and capture potential assessments were conducted
The social, environmental, and economic impact of the use of solar energy systems were determined
Three Proof-of-Concept Experiments were recommended
Phase 1 and 2 development plans were prepared
A utilization plan for implementation and commercialization of solar energy systems was prepared.

FIGURE 8.1: REGIONAL CLIMATIC CLASSIFICATION FOR THE HEATING SEASON (NOVEMBER–APRIL)

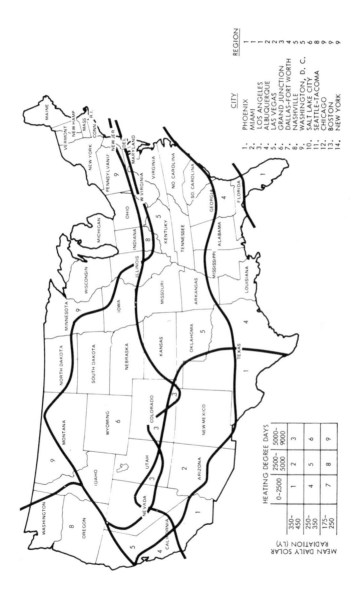

FIGURE 8.2:   REGIONAL CLIMATIC CLASSIFICATION FOR THE COOLING SEASON (MAY-OCTOBER)

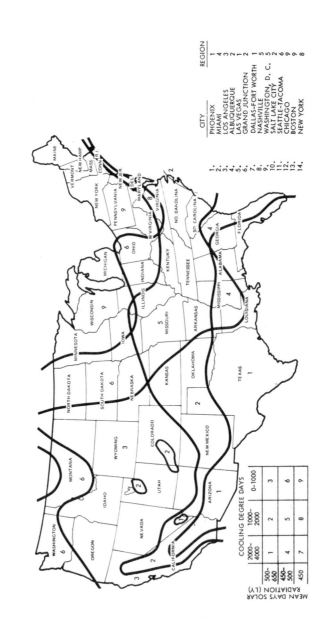

CITY	REGION
1.	PHOENIX	1
2.	MIAMI	4
3.	LOS ANGELES	3
4.	ALBUQUERQUE	2
5.	LAS VEGAS	1
6.	GRAND JUNCTION	2
7.	DALLAS-FORT WORTH	1
8.	NASHVILLE	5
9.	WASHINGTON, D. C.	5
10.	SALT LAKE CITY	2
11.	SEATTLE-TACOMA	6
12.	CHICAGO	9
13.	BOSTON	9
14.	NEW YORK	8

Climatic Regional Classifications

The country was divided into representative heating and cooling regions in terms
of average daily solar insolation and degree-days for both the heating and cooling seasons. Nine climatic regions were identified for each season. The matrices
defining each climatic region and the geographic boundaries of these regions are
shown on Figures 8.1 and 8.2.

Building Types

Using the U.S. Department of Commerce *Construction Review* classifications,
Standard Metropolitan Statistical Area data, and the *Standard Industrial Classification Manual,* a total of 14 building types were initially identified. These
were further reduced to five which represented up to 83% of the total building
construction market and as a group had the highest capture potential. They are
characterized typically by their respective floor areas. The building types selected were

Residential: (1) Single family detached residence—1,400 sq ft
(2) Eight unit apartment building—6,400 sq ft (800 sq ft/unit)
Commercial: (3) Two-story office building—10,000 sq ft
(4) Shopping center—15,000 sq ft
Educational: (5) Elementary school—9,600 sq ft

The above building types were used to establish the heating, cooling, and year-
round domestic hot water loads.

Selected Cities

Fourteen cities with significantly large populations were selected for evaluation.
Additional criteria for their selection included availability of solar insolation
data, building starts, fuel consumption, fuel availability, and per capita energy
consumption projections for 1980, 1990 and 2000. They encompassed only
eight of the nine defined climatic regions since it was determined that there was
no area in the U.S. which satisfied the requirements of Region 7. The 1970
census data were used to identify large population centers. An effort was made
to select cities which were not adjacent to the regional boundary lines, to cover
as much of the U.S. as possible and to include two or three cities in the densely
populated areas. Pertinent data regarding these cities is provided in Table 8.1.

Load Determination

The five selected building types in the fourteen selected cities were used to determine the space heating, space cooling and dehumidification and domestic service
hot water heating loads. In calculating hot water heating energy requirements,
city water temperatures were identified for each month of the year together
with individual building type consumption patterns and a desired temperature
of 140°F.

TABLE 8.1: PERTINENT DATA FOR SELECTED CITIES

No.	Name of City	Heating Season Climatic Region	Cooling Season Climatic Region	Altitude (Ft above Sea Level)	Latitude (in Degrees)	1970 Population (SMSA)	City Rank
1	Phoenix	1	1	1117	33	968,487	34
2	Miami	1	4	7	25	1,267,792	25
3	Los Angeles	1	3	99	34	7,032,075	2
4	Albuquerque	2	2	5310	35	315,774	98
5	Las Vegas, Nevada	2	1	2162	36	273,288	117
6	Grand Junction,Col.	3	2	4849	39	20,170	—
	(Denver)	(3)	(2)	(5280)	(39)	(1,227,529)	(27)
7	Ft. Worth	4	1	544	32	762,086	43
8	Nashville	5	5	577	36	540,982	61
9	Washington, D.C.	5	5	14	39	2,861,123	7
10	Salt Lake City	6	2	4220	40	557,635	58
11	Seattle	8	6	386	47	1,421,869	17
12	Chicago	9	9	658	42	6,978,947	3
13	Boston	9	9	15	42	2,753,700	8
14	New York	9	9	19	40	11,528,648	1

For calculating space heating, space cooling and dehumidification loads, indoor design criteria of 68°F were used for the winter season and 78°F (50% RH) for the summer season. Coefficients of transmission for the building envelopes were selected with a view towards conserving energy. Design assumptions for interior heat gains and ventilation requirements were based on recommendations in the ASHRAE *Handbook of Fundamentals* and the Chicago Code.

Space heating loads incorporated the use of a thermal recovery unit. The calculations of space heating, space cooling and dehumidification loads are typical thermal loads rather than peak loads. Average monthly maximum/minimum ambient temperature values were distributed over a diurnal swing to produce design values for eight three-hour time intervals to represent the conditions for a typical day for each month of the year. For the cooling season, coincident values of wet and dry bulb temperatures were used to determine change in enthalpy. Annual loads for the selected building types and cities are shown in Table 8.2.

TABLE 8.2: SUMMARY OF ANNUAL SPACE HEATING AND COOLING LOADS (Btu x 10^3)

No.	City	Building Type	For Space Heating	For Space Cooling	Total Annual
1.	Phoenix	Store	0	1,792,850	1,792,850
		Elementary School	70,705	516,870	587,575
		Apt. House	66,376	190,592	256,968
		Single Family Det. Res.	20,877	64,900	85,977
		Office Building	34,825	546,330	581,155
2.	Miami	Store	0	2,773,910	2,773,910
		Elementary School	9,360	792,110	801,470
		Apt. House	2,184	274,880	277,064
		Single Family Det. Res.	939	92,560	93,499
		Office Building	4	855,320	855,324
3.	Los Angeles	Store	0	705,292	705,292
		Elementary School	64,055	149,903	213,958
		Apt. House	57,816	40,200	98,016
		Single Family Det. Res.	18,397	18,353	36,750
		Office Building	18,285	139,072	157,357
4.	Albuquerque	Store	0	881,681	881,875
		Elementary School	143,619	245,997	389,616
		Apt. House	168,384	81,232	249,616
		Single Family Det. Res.	51,971	29,402	81,373
		Office Building	89,935	252,554	342,699
5.	Las Vegas, Nev.	Store	0	1,445,716	1,445,716
		Elementary School	85,479	415,408	500,887
		Apt. House	94,928	157,016	251,944
		Single Family Det. Res.	29,422	53,606	83,028
		Office Building	45,491	436,989	482,480
6.	Denver, Colo.	Store	6,993	606,382	609,375
		Elementary School	212,995	171,974	384,969
		Apt. House	244,880	50,008	374,934
		Single Family Det. Res.	75,589	19,326	94,915
		Office Building	145,840	168,772	314,612

(continued)

TABLE 8.2: (continued)

No.	City	Building Type	For Space Heating	For Space Cooling	Total Annual
7.	Ft. Worth	Store	0	1,553,745	1,553,745
		Elementary School	73,162	429,112	502,274
		Apt. House	83,984	158,200	242,184
		Single Family Det. Res.	26,008	54,183	80,191
		Office Building	30,069	459,974	490,043
8.	Nashville	Store	0	1,188,383	1,188,383
		Elementary School	110,867	344,234	445,101
		Apt. House	134,536	118,800	253,336
		Single Family Det. Res.	126,536	41,384	82,9)
		Office Building	63,852	364,414	428,266
9.	Washington, D. C.	Store	0	943,973	943,973
		Elementary School	112,117	267,194	379,311
		Apt. House	161,784	87,200	248,984
		Single Family Det. Res.	49,892	31,566	.81,458
		Office Building	78,866	276,715	355,581
10.	Salt Lake City	Store	5,828	685,085	690,913
		Elementary School	202,620	199,455	408,075
		Apt. House	237,160	61,976	335,168
		Single Family Det. Res.	73,198	22,900	96,098
		Office Building	138,531	202,836	341,367
11.	Seattle	Store	0	307,709	307,709
		Elementary School	158,569	68,212	226,781
		Apt. House	201,592	16,384	217,913
		Single Family Det. Res.	62,359	8,179	70,538
		Office Building	87,712	63,446	151,158
12.	Boston	Store	2,883	551,999	554,832
		Elementary School	175,994	143,015	319,009
		Apt. House	219,440	45,536	264,976
		Single Family Det. Res.	67,598	17,473	85,071
		Office Building	115,192	143,415	258,607
13.	Chicago	Store	7,127	674,985	682,112
		Elementary School	190,637	187,465	378,102
		Apt. House	245,216	60,448	295,664
		Single Family Det. Res.	72,425	22,260	94,685
		Office Building	135,609	192,000	327,609
14.	New York	Store	130	774,801	774,931
		Elementary School	149,550	215,444	364,994
		Apt. House	188,760	69,104	257,864
		Single Family Det. Res.	58,228	25,375	83,603
		Office Building	93,346	219,098	312,444

REFERENCE SYSTEM DESIGNS

Relying on existing and proven solar and conventional climatic control hardware, three reference designs were selected for detailed cost evaluation in each of the fourteen cities and five building types. The systems were designated Reference System A for domestic hot water system, Reference System B for hot water and space heating, and Reference System C for hot water, space heating and space cooling. For all three reference systems glazed metallic collectors employing water or water/ethylene glycol mixtures were used. For the space cooling function, hot-water-fired lithium-bromide gas absorption refrigeration system was employed.

Mathematical models were constructed of the Reference Systems, and the resulting quantitative data generated resulted in some changes in the systems components, controls, and operation modes to enhance performance and reduce initial and operating costs.

Hot Water (System A)

The reference system (System A) for furnishing hot water only is shown in Figure 8.3. Operation is initiated with the flow of water into the collector, SC-1.

FIGURE 8.3: HOT WATER SYSTEM (SYSTEM A)

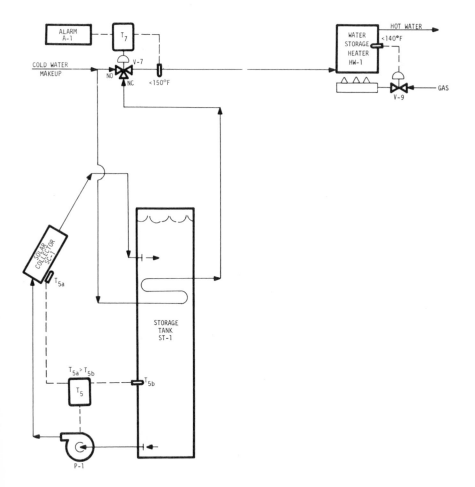

If the temperature of the fluid, which is a mixture of water and ethylene glycol, in the collector exceeds that of the fluid in the storage tank, **ST-1**, the fluid will flow from the collector to the storage tank with circulation provided by pump, **P-1**. If, however, the temperature of the collector is less than that of storage, the pump is inactivated. This is implemented by temperature sensors T_{5a} and T_{5b} which control the motor of the pump, **P-1**.

As hot water is taken from the system, make-up cold water will flow through the control valve, **V-7**, either from the finned coil in the storage tank or directly through to the water storage heater unit, **HW-1**. The temperature of the cold water is sensed on the output side of the valve, at T_7, causing the valve to close, thus forcing the water to flow through the finned coil in the storage tank to absorb heat. When the fluid becomes excessively heated during periods of high insolation, the temperature of the fluid leaving the heat exchanger coil might exceed $150°F$. When this occurs, the sensor T_7 causes the valve, **V-7**, to intro- duce cold water until the temperature falls below $150°F$. Every time valve **V-7** introduces cold water in this way, it turns on the alarm, **A-1**. The purpose of this is to provide an indication of the event that valve **V-7** is locked in this posi- tion.

The water from either source is then conducted to a standard hot water heater, **HW-1**. It maintains water at temperatures of at least $140 \pm 5°F$ through the use of auxiliary power.

Hot Water and Space Heating (System B)

The selected reference system (System B) for supplying hot water and space heating to buildings is shown in Figure 8.4. The part of the system necessary to supply hot water, i.e., the collector, storage, and the hot water storage and heat unit, is basically the same as the Reference A Hot Water System. However, the sizes of some components, and their performance characteristics, would change because of the additional requirements imposed by the space heating load. The space heating portion of this system operates, to satisfy thermostat demands, in one of three automatically selected modes.

Solar Heating: This is the normal mode, using solar power only. Hot fluid from either the collector or the storage tank furnishes heat to the preheat **PC-1**.

Auxiliary Boost: This is the mode in which part of the heat is supplied from solar sources by the preheat coil, **PC-1**, and the remainder of the required heating is furnished by the heating coil, **HC-1**, using the auxiliary heater, **HW-2**.

Auxiliary Heating: This is the mode in which all the heating is accomplished by means of the heating coil, **HC-1**, through the use of the auxiliary heater, **HW-2**, which is set to produce hot water at $120°F$. In this mode, the temper- ature of the fluid from either the collector or the storage tank is below some specified value, e.g., $70°F$. The flow of this fluid is stopped by turning off the motor driving the pump, **P-3**.

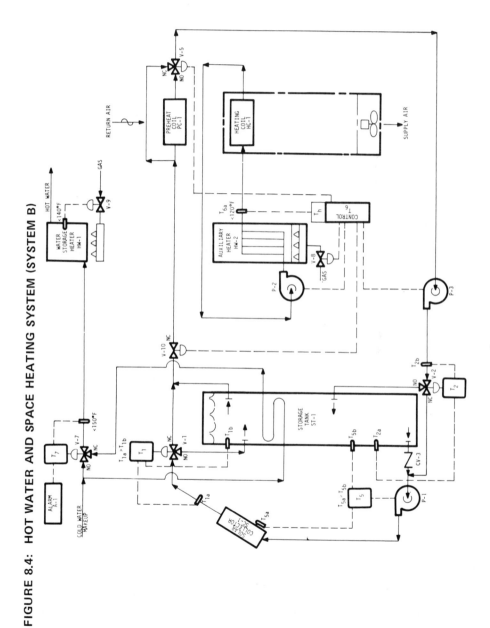

FIGURE 8.4: HOT WATER AND SPACE HEATING SYSTEM (SYSTEM B)

Controls are of the bang-bang type, i.e., they are either fully on for as long as required, or off completely.

Combined Hot Water, Space Heating and Cooling System (System C)

The selected Reference System (System C) to furnish the combined functions of hot water heating, space heating and cooling is shown in Figure 8.5. The system operates in either the heating or cooling mode as manually selected through the control unit. The system furnishes hot water as required irrespective of the mode selected.

Signals from the control unit will connect components as they exist in the hot water and space heating portion of the reference system. If the fluid flowing from the auxiliary heater, HW-2, leads to excessively hot air from the duct, control of the heat input to HW-2 will be switched to a second Aquastat set at a lower temperature, e.g., $160°F$. The system operates in the cooling mode in two ways, i.e., either the air conditioner is powered entirely by the solar energy system, or completely from the auxiliary heater.

The system when in the auxiliary powered mode operates in the following manner. A signal from the control unit will change the position of valve, V-6, so that the fluid can flow (1) through the pump P-2, (2) through the auxiliary heater, HW-2, (whose temperature is controlled at $210°F$ by means of a standard Aquastat), (3) through the coils of the generator, G-1, (4) through the valve, V-3 (which will be changed in position by temperature measurements made by sensors T_{3a} and T_{3b} and controller, T_6), and (5) back to valve, V-6. Cooling of the return air from the building, circulated by a fan, is accomplished in the air conditioner, AC-1, in the following way:

> The generator, G-1, provided with heat from the fluid, causes the refrigerant, water, to evaporate at low pressure (66 mm of mercury, for instance), and compresses the vapor.
>
> The steam is condensed in condenser, C-1, at low temperature (e.g., $90°F$), and the resulting liquid water is allowed to enter the cooling coil, CC-1, through an orifice, at a lower pressure (6 mm of mercury).
>
> The warm air in the air stream causes the water in the cooling coil, CC-1, to evaporate, and thus cool the building.
>
> The steam at low pressure is absorbed by the lithium bromide and cooled (e.g., at $75°F$).
>
> After exchanging heat this mixture is recirculated to the generator.

The heat rejected from the condenser, C-1, and the absorber, must now be exhausted to the external atmosphere. This is accomplished by circulating cool water through a coil around the absorber and then another coil around the condenser which is forced to flow to a cooling tower, CT-1, by means of a pump, P-4.

FIGURE 8.5: HOT WATER, SPACE HEATING AND COOLING SYSTEM (SYSTEM C)

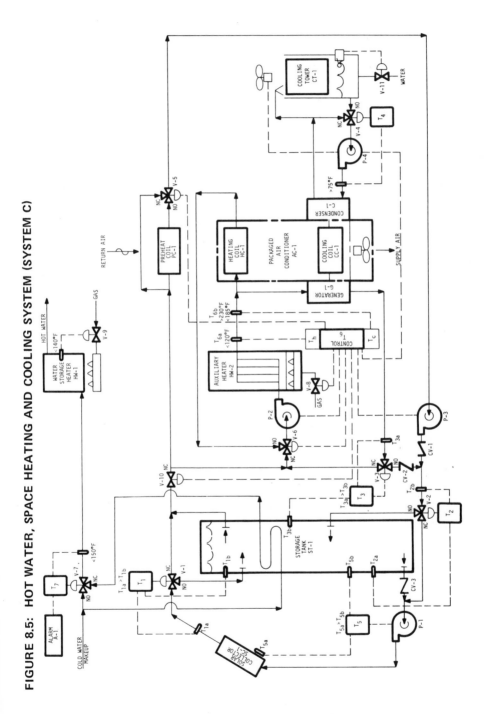

The cooling tower, **CT-1**, cools the water by the action of both evaporation and counterflow forced air convection. The cooled water collects in a sump at the base of the cooling tower, **CT-1**. The water lost through evaporation is replenished from the main water supply automatically by means of a level sensor controlling a valve. The water leaving the cooling tower will be at the ambient wet-bulb temperature, or somewhat higher, depending on the cooling load (approximately $10°F$).

The temperature of the coolant and the fluid supplied to the generator determine the maximum cooling rate that can be supplied under these conditions. The heating input that must be supplied to the generator is equal to the required cooling load divided by the coefficient of performance, which is 0.65. Because of the considerable mass in the generator structure and associated hardware in the air conditioner, a lag of 20 minutes at start-up will follow the thermostat signal before actual cool air becomes available.

Computer Simulation of Reference Systems

The general approach was to represent each system as a network analog, similar to a circuit diagram in electrical analysis. A single program was prepared which simulates any of these Reference Systems. It contains three sections which model the Hot Water System, the Hot Water and Space Heating, and the Hot Water and Space Heating or Cooling Systems. By combining these sections (depending on whether or not a demand exists) the overall system performance is determined.

A matrix of system attributes and components/functions was developed on a format for considering the operational requirements for the various reference system configurations. The principal attributes which were analyzed for each component included reliability, durability, maintainability, and life safety. A failure modes and effect analysis were performed for the collector. Components were ranked with respect to frequency of repair. The three reference systems were shown to be quite serviceable and well-adapted to the specific cost/performance needs. They are operational over 99% of the time, repairable in about 2 hours when there is a failure, and even considering the possibility of nighttime or weekend failures, the mean down time was in the order of a day for the more complex reference system designs.

Cost Analysis

The cost analysis takes as inputs the climatic conditions and loads for the various building types in each city and through a series of computer programs determines system cost and percent solar utilization for each of the Reference Systems. The market capture analysis was integrated into the cost programs since these tasks were interrelated. The market penetration analysis was a natural outgrowth of the cost program because the inputs to that analysis were direct outputs of the cost program. The basic output were sets of curves showing the capital costs as a function of system utilization and cumulative costs as related to years of operation.

From the results of the cost analyses it became evident that solar systems have a clear cost advantage for hot water heating when compared to conventional systems using electricity for commercial buildings. The solar system is at a disadvantage for the single family residence largely because of the preferential electrical rate structure currently given to them. However, if the rate structure for new single family residences is brought into line with commercial building rates then the solar hot water heating system will also become cost competitive in this extremely important area. For an 8-unit apartment building the hot water heating system would be cost effective today and even more so in 1980.

Hot water heating is the particular functional requirement where solar energy shows its greatest relative cost advantage. The load is relatively constant and the solar system can be made small and highly effective throughout the year. Only in cities where electricity costs are low (i.e., Seattle, Nashville, and Las Vegas) due to the availability of hydroelectric power, does the conventional system show a long-term cost advantage over the solar energy system.

The solar climate control system for hot water and space heating shows an advantage for most apartments, schools, and a few office buildings. However, if the heat generation in the building becomes large compared to heat losses, heating requirement occurs only during the coldest months of the year. This is not compatible with the low solar insolation during the heating season and hence the incremental cost for satisfying these loads is high. Therefore, the solar space heating system becomes less competitive under these circumstances.

Thus, solar heating systems for some office buildings and most shopping centers become more expensive than the conventional systems because of the sizable internal heat generation. For the single family residence, the solar space heating is penalized by the preferential electric heating rates and is currently more costly than the conventional systems. A change in this rate structure could alter the costs in favor of the solar heating system.

A second factor that reduces the cost effectiveness of the solar space heating system for single family residences is the high unit cost of the system since the cost of the auxiliary equipment is prorated over a comparatively small collector area (in comparison to apartments and commercial buildings). This disadvantage would disappear for a larger single family residence. In this study, a 1,400-ft2 residence with energy conserving construction was chosen as an example. Since many new residences are larger than 1,400-ft2 or may not employ the energy conserving construction techniques assumed for this building type, the solar energy space heating system would become more cost effective.

The complete solar climate control system using the absorption refrigeration system (Reference System C) has a cost advantage over conventional systems only if government incentives are available or major improvements in technology of the absorption refrigeration unit can be achieved. The high temperature (180° to 210°F) requirement for these cooling units reduce the performance of the solar energy system while increasing costs substantially.

In addition, the much higher coefficient of performance of the mechanical compression refrigeration systems compared to the absorption systems results in a considerable cost leverage for the amount of electrical energy for these conventional cooling units. Furthermore, the absorption system usually requires a water tower for heat rejection, which is more expensive than a comparable air cooled mechanical unit. Therefore, it will be necessary to considerably improve the absorption unit performance through selecting cycles or materials to improve the COP, or reduce the required operating temperatures, before the overall solar heating and cooling system can be cost-competitive with conventional systems.

CAPTURE POTENTIAL ASSESSMENT

Case Assumptions

Four alternate cases were selected to illustrate relative SES (Solar Energy System) market performance, given varying economic and government role assumptions. These assumptions reflect a determination of economic factors which will have a strong influence on SES market performance: capital cost, fuel price and availability, and potential government financial participation.

Case I, the base case, is the most conservative estimate of possible SES market performance. It is based on detailed cost estimates and on relatively conservative estimates of fuel price escalation (see Table 8.3). These fuel price estimates are based on a model embodying the important cost parameters of the electric and gas utility industry. The base case also assumes the current electric utility practice of preferential rates for single-family homes and a slight moderation of the recent increasing trend toward electric heating. No direct government incentives to encourage SES purchase are assumed. The SES industry is presumed to be in essentially a market introduction stage.

Furthermore, no cost credit is given for the fact that the collector will be integrated into the roof in new construction, resulting in a portion of the cost actually being new construction cost and not incremental SES cost.

Case II assumes the same cost and fuel price escalation as the base case, but also assumes direct government incentives of a 25% tax credit on the SES incremental mortgage payment to SFR consumers, and an investment tax credit of 7% for MF and commercial markets.

Case III assumes the same system cost, fuel availability, rate structure, and government role as the base case, but assumes a 1980 system cost reduction of 25%. This case would reflect a large government- or industry-initiated R&D effort, and a more advanced SES industry state-of-the-art. It also could reflect part of the SES cost being absorbed into the construction cost of the building. This case illustrates SES relative market sensitivity to capital cost.

TABLE 8.3: FORECASTED ENERGY RATE LEVELS

Residential Gas Rate Level

Region	1973	1980	1985	1990	2000
East	1.00	1.57	2.11	2.21	2.50
South East	1.00	1.50	1.99	2.33	3.20
East Central	1.00	1.67	2.39	2.67	3.50
South Central	1.00	2.03	3.06	3.45	4.20
West Central	1.00	1.66	2.31	2.50	2.90
West	1.00	1.67	2.34	2.62	3.40

Electric Rate Level

Region	1973	1980	1985	1990	2000
East	1.00	1.37	1.55	1.7	2.2
South East	1.00	1.61	1.87	2.1	2.8
East Central	1.00	1.34	1.55	2.0	2.4
South Central	1.00	1.66	2.07	2.4	3.1
West Central	1.00	1.35	1.65	2.0	2.7
West	1.00	1.43	1.68	1.9	2.6

Commercial Gas Rate Level

Region	1973	1980	1985	1990	2000
East	1.00	1.66	2.31	2.50	2.70
South East	1.00	1.86	2.71	3.00	3.61
East Central	1.00	1.80	2.66	2.95	3.60
South Central	1.00	2.22	3.44	3.69	4.10
West Central	1.00	1.85	2.69	3.10	4.03
West	1.00	1.86	2.72	3.15	3.55

Oil Rate Level

Region	4Q 1973	1980	1985	1990	2000
East	1.00	1.70	1.87	2.06	2.40
South East	1.00	2.00	2.20	2.42	2.84
East Central	1.00	1.85	2.05	2.25	2.63
South Central	1.00	2.00	2.20	2.42	2.84
West Central	1.00	1.85	2.05	2.25	2.63
West	1.00	1.85	2.05	2.25	2.63

Case IV assumes the same system cost, fuel escalation, and government role as the base case, but also assumes that the preferential rate structure for single-family residences is abolished and reestablished at multifamily rates. This case illustrates the sensitivity of SES market penetration to fuel cost or rate changes, and will give an indication of probable effects of fuel price escalations above those estimates.

Energy Supply and Demand

Figure 8.6 compares four energy demand scenarios with the available domestic supply. Assumptions underlying each scenario are shown on the following page.

FIGURE 8.6: DEMAND SCENARIOS COMPARED WITH ENERGY SUPPLY

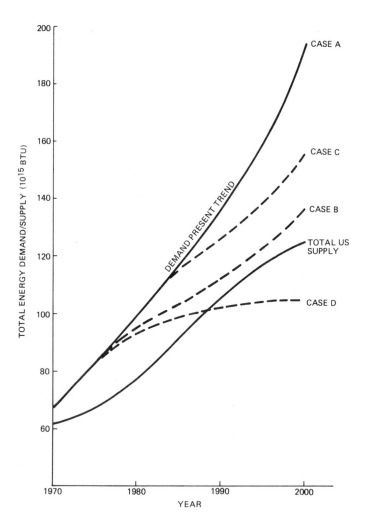

Case A: Continuation of present population, energy consumption, and GNP growth rates. Fuel costs will rise with the general price level.

Case B: The annual population growth rate will decline to zero by 2050, growth in real GNP will decline to 2.5% by 2000, conservation techniques will increase energy savings 18% by 2000, and the price of energy will increase by a factor of 3.2 by the year 2000, but will not affect consumption.

Case C: Same as Case B except impact cannot start until 1985. Due to long range planning, electric power generation will be capable of meeting demand incorporated into Case A using nuclear energy.

Case D: Same as Case B except the price of energy will increase by a factor of 4 above the 1970 base in the year 2000, and will reduce consumption by 25%.

If a stabilized energy economy can be developed as shown in Case D, there will be no shortage or need for imported fuels after 1990. Case C is perhaps the most realistic of the scenarios. Under this scenario the difference between the supply and demand remains relatively constant after 1985.

Capture Potential Forecast

Base Case Results: Detailed results of the base case analysis are presented in Table 8.4. Major summary points are as follows. Hot water systems are substantially more competitive than space heating systems for all building types and all regions. This reflects the lower costs for hot water systems, as well as the stability of the hot water load over the year.

The multifamily, low-rise apartment market is the most advantageous market for SES. Fairly large capture (12% for hot water heating) occurs as early as 1980.

Major reasons for the feasibility of this market include high electric fuel prices, moderate internal/external heat load, relatively low unit cost (compared to costs for other building types), low heat loss (fewer walls exposed to the outdoors), high hot water and heating load as a percent of total load, and economies of scale (eight units as opposed to a single unit in a SFR, for example).

Furthermore, the number of building starts and the percent electric fuel penetration are relatively high. Schools have approximately the same performance advantages that apartments do, but because fuel costs are less, and building starts and electric fuel penetration are low, schools do not generate so large a total market.

TABLE 8.4: TOTAL SOLAR ENERGY SYSTEM MARKET, BASE CASE, 1980 to 2000

Building Type	System	1980				1990				2000			
		Units	% Market Capture	$ (000)	Sq.Ft (000)	Units	% Market Capture	$ (000)	Sq.Ft (000)	Units	% Market Capture	$ (000)	Sq.Ft (000)
Single-Family Residence	HW	11,240	.98	19,470	875	36,435	3.4	50,698	2,760	49,585	4.5	67,128	3,835
	Space	1,430	.12	7,641	274	3,775	.35	17,631	711	6,700	.61	29,138	1,291
Multi-Family Low-Rise	HW	14,490*	12.0	59,461	3,694	26,230	22.5	93,422	6,613	31,670	25.6	103,458	8,000
	Space	6,520	5.4	64,829	3,095	15,270	13.2	132,840	7,256	19,530	15.8	154,488	9,232
Schools	HW	540*	7.4	1,929	102	1,000	14.3	3,057	184	1,540	22.0	4,344	287
	Space	225	3.1	2,260	105	525	7.6	4,457	236	895	12.5	6,938	402
Commercial	HW	650	.87	1,481	68	1,140	1.6	2,305	120	1,790	2.4	3,316	188
Total New Construction		35,095	-	157,072	8,213	84,375	-	304,410	17,800	111,710	-	368,808	23,235
Retrofit													
Single-Family Residence	HW	5,650	-	14,700	441	23,000	-	48,817	1,794	34,500	-	65,546	2,691
Multi-Family Low-Rise	HW	3,900	-	24,000	995	14,000	-	71,869	3,570	21,000	-	94,325	5,355
	Space	2,000	-	29,850	950	3,500	-	43,857	1,663	6,500	-	70,821	3,088
Schools	HW	1,000	-	5,358	189	4,700	-	20,524	888	9,500	-	37,160	1,796
	Space	320	-	4,824	150	1,600	-	19,658	747	3,200	-	35,195	1,494
TOTAL		47,965	-	235,804	10,938	131,175	-	509,135	26,542	186,410	-	671,855	37,659

*Units actually refer to number of installations. Since the building type is an 8-unit apartment, each installation reflects capture of 8 apartment units.

Initial penetrations for apartments are high, but do realistically reflect performance in relationship to performance requirements. Return on investment is as high as 40% in 1980 when SESs are compared to electric usage, and in some cases rises to as high as 75% by 2000.

Commercial buildings do not represent large markets for SESs. Reasons include low heating and hot water loads, relatively low fuel prices and high system costs, high internal to external load, and low electricity usage percent penetrations. Furthermore, builders/developers are first-cost sensitive, are not highly concerned with operating costs and life-cycle costing, and are generally resistant to technological innovation.

Markets increase fairly rapidly between 1980 to 1990, but experience a steadying growth between 1900 and 2000. Reasons reflect marketing consideration regarding product life-cycle, and relative fuel price/system cost escalations. The increased capture between 1990 and 2000 is due to increased gas market penetration (by 2000, gas prices in many SMSAs are equal to electric prices) and to marketing factors reflecting higher acceptance of SES. By the year 2000, approximately 20% of the new residential construction market can be captured by SES. Notice that about 85% of this market will be in multifamily low-rise structures.

The retrofit market reflects substantially lower capture rates than the new construction market, but because the base is so high, generates a significant additional market. Retrofit market feasibility is low because financing charges are higher, system cost is generally 50 to 100% higher, and system efficiency is lower than for new construction.

Base Case Results: The regional markets for SES are summarized in Table 8.5. The SES industry will be local market and not a regional market, but comparisons among regions do indicate areas of relative potential. The Southern region represents the largest market in all decades, with the Western region the second largest in 1990 and 2000.

The decline in market share for the Northeast region reflects decreased building construction and the lower rate of fuel price escalation. The high initial market reflects the high fuel prices in the Northeast region and the relatively high usage of oil rather than gas as an alternate to electricity. This high market in the Northwest region illustrates the importance of fuel price variations compared to climatological variations among regions. The Western and Southern regions have climate as well as fuel price advantages.

Alternate Case Results: SES market potential under the alternate cases is summarized in Figures 8.7 and 8.8. Case III (25% system cost reduction) results in the largest potential unit market. A reduction of 25% in the cost increases the total unit market by approximately 50%. The percentage increase in the market, given a specific cost reduction, will depend on the building type current position on the market penetration curve, reflected in existing market share.

TABLE 8.5: REGIONAL SOLAR ENERGY SYSTEM MARKET, BASE CASE, 1980 to 2000

	1980			1990			2000		
	$ (000)	Sq.Feet (000)	Sq.Ft. % of Total	$ (000)	Sq.Feet (000)	Sq.Ft. % of Total	$ (000)	Sq.Feet (000)	Sq.Ft. % of Total
West	32,744	1,631	20%	97,465	5,454	30%	115,395	6,943	30%
North East	42,806	2,098	26%	59,510	3,277	19%	74,755	4,426	19%
South	60,874	3,495	42%	113,695	7,335	41%	133,330	9,246	40%
Central	20,648	989	12%	33,740	1,814	10%	45,330	2,620	11%
Total U.S.	157,072	8,213	100%	304,410	17,880	100%	368,810	23,235	100%

FIGURE 8.7: SOLAR ENERGY SYSTEM MARKET IN DOLLARS

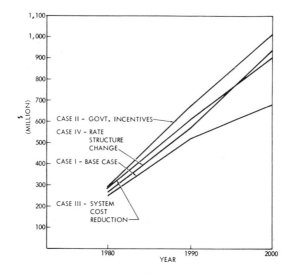

FIGURE 8.8: SOLAR ENERGY SYSTEM MARKET IN SQUARE FEET

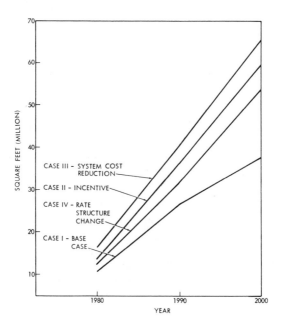

For example, the SFR market will be on the relatively flat part of the normal curve in 1980 (only 0.98% penetration), and thus will increase more slowly, given the same cost reduction, than the apartment market which is in the steeper section of the curve (12% penetration in 1980).

The largest dollar market is generated by Case II (incentives case). Case IV (SFR rate structure change) actually results in the largest direct dollar and unit increase, but because it applies only to SFR, the effect on total market growth is not so large as in the other cases. However, this case resulted in the largest increase for the SFR market.

For example, a system cost reduction of 25% (Case III) resulted in a hot water heating system penetration of 1.67% in 1980 (compared to 0.98% in the base case), but the rate structure change resulted in a capture of 2.05%. Results were similar for all decades and both reference systems. In fact, by the year 2000, this case resulted in an 11.2% market capture for hot water systems (compared to 4.5% in the base case) in the SFR market. This is a significant increase, and one that has a fairly good probability of actually occurring.

Energy Savings Implications: Yearly energy savings by the year 2000 under each case assumption are summarized in Table 8.6.

TABLE 8.6: SES MARKET CAPTURE POTENTIAL AND ENERGY SAVINGS FOR THE YEAR 2000

Case	Area (Million) (sq-ft/yr)	Cumulative Area (Million) (sq-ft)	Market ($Million/yr)	Yearly Energy Savings (10^{14} Btu)
I	38	447	680	1.29
II	60	720	1010	2.09
III	65	812	900	2.36
IV	54	635	940	1.84

For solar to satisfy 1% of the total nation's energy requirements under energy demand scenario C, the following situation would have to occur. The probabilities listed assume the fuel rate projections are correct. More rapid fuel price escalation would greatly increase the probability of this situation occurring, and would permit an increase in the required dollars per square foot.

Requirement	Comments	Probability of Occurrence
Capture for SFR new construction would be 55% for space heating systems.	Requires a system cost reduction to approximately $2.00/ft^2 installed. This would require massive R&D effort and government incentives.	Possible
Capture rate for MF new construction would be 55% for space heating systems.	Requires a system cost reduction to approximately $3.50/ft^2 installed. This would require a massive R&D effort and government incentives.	Possible
Retrofit applications would be about 40% the total square footage. This would require a 1% capture of the total retrofit market (uniform 1%, can be higher for some building types and lower for others).	Current projections show retrofit applications at 40% of the total. However, % capture is low. At required cost reductions, higher penetrations are possible.	Probable
Capture rate for commercial new construction would be 10%.	Case III (25% system cost reduction) reflects 4.1% capture for hot water heating only. Lower system costs will make space heating feasible, thus increasing capture.	Highly Probable
Capture rate for institutional new construction would be 50% for space heating systems.	Case III reflects a capture of 22% for space heating. Required cost reductions would result in 50% capture, but because of the low number of starts, the overall impact is not so great.	Highly Probable

Notice that the above scenario reflects solar utilization for hot water heating and space heating only. If solar can be made competitive for cooling, the required capture rates would decrease by one-third, since overall energy savings for each building type would be greater. Larger energy savings potential for SES would be indicated given a competitive system for hot water heating, space heating and space cooling.

PROOF-OF-CONCEPT EXPERIMENTS

Reference System Summary

The initial Phase 0 study resulted in the selection of three reference systems. These were designated System A (Hot Water System), System B (Hot Water and Space Heating), and System C (Hot Water, Space Heating and Cooling) and were used to establish capture potential for solar energy systems. While these reference systems were cost-effective for many building types and cities, it was apparent that new, innovative designs should also be evaluated to further improve their cost-effectiveness, particularly for the overall combined hot water, space heating and cooling systems.

The results of these evaluations led to the selection of three solar augmented heat pump systems as a means of reducing both initial costs as well as operating costs. The three solar augmented heat pump systems were designated Reference Systems D, E, and F. Their general characteristics are given in Table 8.7.

TABLE 8.7: GENERAL CHARACTERISTICS OF SOLAR AUGMENTED HEAT PUMP SYSTEMS

Reference System Designation	Major System Component	System Opération
(D)	• Flat-Plate Solar Collector • Crushed Rock Energy Storage (Hot and Cold) • Variable Pressure Ratio Heat Pump • Auxiliary Hot Water Heater • Collector Air Circulation Fan	• All air circulating system (collector and distribution loops) • Heat pumps with "reverse-air" or "reverse-refrigerant" options available
(E)	• Flat-Plate Solar Collector • Water Storage Tanks (Hot and Cold) • Variable Pressure Ratio Heat Pump • Auxiliary Hot Water Heater • Collector Water Circulation Pump	• Water circulating system (collector loop) • Water or air circulating system (distribution loop) • Heat pump with "reverse-refrigerant" on "reverse water" loop options available

(continued)

TABLE 8.7: (continued)

Reference System Designation	Major System Component	System Operation
(F)	• Flat-Plate Solar Collector • Water Storage Tanks (Hot and Cold) • Conventional Heat Pump (with Evaporative Water-Cooled Condenser) • Rock-Bed Regenerator • Auxiliary Hot Water Heater • Collector Water Circulation Pump • RBR Air Circulation Fan	• Water circulating system (collector loop) • Water and air circulating systems (distribution loop) • Heat pump with "reverse-water" loop only • RBR used to cool or regeneratively heat air for gym space • Collector loop used to heat office space • Heat pump loop used to cool or heat office space

All three systems were designed to provide the combined building functions of service for hot water, space heating, and space cooling. The thermal models and system diagrams for Reference Systems D and E are shown in Figures 8.9, 8.10, 8.11 and 8.12.

Systems D and E utilize a combination of solar augmentation and a variable pressure ratio heat pump to achieve significantly higher coefficients of performance than that achievable with a conventional heat pump. System D, which is essentially an all-air system (except for the hot water function), uses crushed rock for both the hot and cold energy storage.

System E circulates water through the collector loop but permits the option of using water or air in the building distribution loop. Both systems utilize heat pumps which are capable of operating with either reverse-refrigerant or reverse-air or water loops.

Another system, F, was also defined. The system, shown in Figure 8.13, is essentially a state-of-the-art modification of System E. It includes a rock-bed regenerator to provide cooling and regenerative heating of circulation air and a conventional heat pump rather than a variable pressure ratio unit which is to be developed concurrently in Phase 1.

FIGURE 8.9: SCHEMATIC DIAGRAM OF REFERENCE SYSTEM D THERMAL MODEL

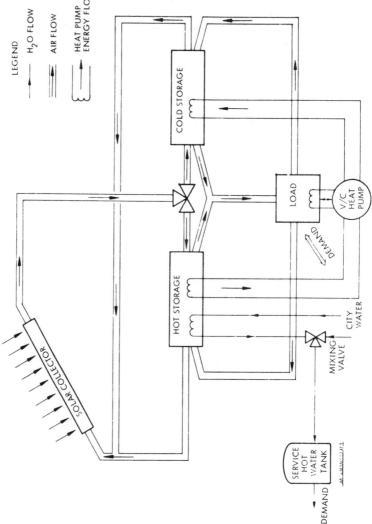

FIGURE 8.10: REFERENCE SYSTEM D SOLAR HEAT PUMP SYSTEM

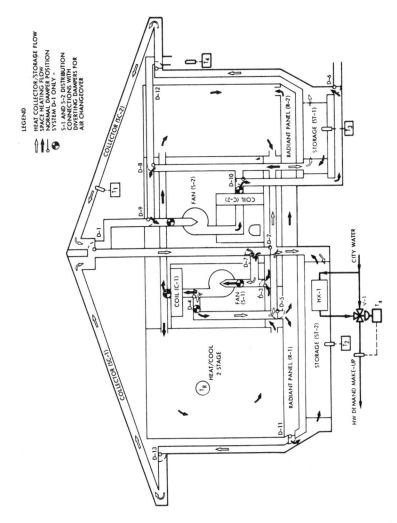

FIGURE 8.11:   SCHEMATIC DIAGRAM OF REFERENCE SYSTEM E THERMAL MODEL

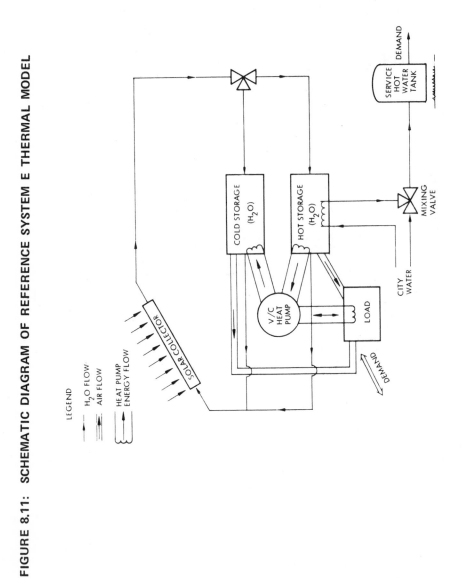

FIGURE 8.12: SCHEMATIC OPERATIONAL DIAGRAM—REFERENCE SYSTEM E

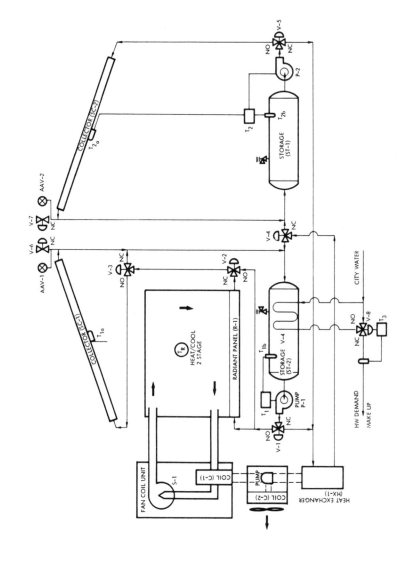

FIGURE 8.13: SCHEMATIC DIAGRAM OF REFERENCE SYSTEM F

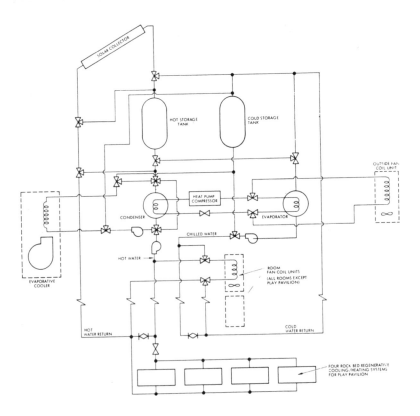

Scottsdale Girls' Club Building

Proposal: The Scottsdale, Arizona, Girls' Club is planning to construct a new building of moderate size (i.e., first floor—12,600 ft², second floor—2,700 ft²). Plans had been completed and construction was to be initiated in September 1974. Discussions with the board members of this public building resulted in their enthusiastic support for the possibility of providing both solar heating and cooling for this facility.

Both Reference Systems D and E were investigated for their applicability to this building. However, Reference System D, which uses air circulation in both the collector and distribution system loop must be architecturally integrated into the building. Since the plans for the Scottsdale Girls' Club building had been completed and approved, the Reference System D was eliminated from consideration.

Reference System E, which uses water circulation in the collector loop and either water or air in the distribution system loop, was determined to be more compatible for incorporation into the existing building plans. However, several factors contributed toward requiring that some modifications be made to Reference System E. These were:

Since the variable pressure ratio heat pump would not be available in time to meet the construction schedule, a conventional heat pump employing reverse-water operation was substituted.

The climatic conditions in Scottsdale permit the use of an evaporatively cooled condenser, and, since this will enhance the COP obtainable from the heat pump, it was substituted for the ambient air-cooled condenser.

The Girls' Club recreational area requires 2000 cfm of outside make-up air to meet ASHRAE standards but dehumidification is not essential, so a rock-bed regenerator to provide summer cooling and regeneration of the make-up air for winter heating was included to reduce the heat pump loads and conserve on the use of electrical energy.

Thus, the modified system, designated Reference System F, was decided upon. The system design characteristics are given in Table 8.8.

TABLE 8.8: SCOTTSDALE GIRLS' CLUB BUILDING DESIGN CHARACTERISTICS

Quantity	Subsystem Description	Design Characteristics
1	Solar Collector (and Circulating Water Pump)	1000 sq. ft. Single-Glaze Fluid-Water Flow Rate-5000 gals/hr
1	Energy Storage Tank (Hot)	2000 Gallons Fluid-Water
1	Energy Storage Tank (Cold)	2000 Gallons Fluid-Water
1	Heat Pump	20 Ton Capacity "Reverse-Water" Operation Evaporative Water-Cooled Condenser
4	Rock-Bed Regenerator	2000 CFM Ventilation Flow Cool & Ventilate Recreational Area Only Pre-heat Ventilation Air for Heating Mode
4	Air Circulation Fan	2000 CFM 0.65 in. H_2O, Gage
1	Auxiliary Hot Water Tank (and Circulating Water Pump)	100 Gallons Electrically Heated

Rationale for Selection: The Scottsdale Girls' Club building was selected because it was compatible with Phase 1 Development Plan ground rules. The funds for the building will be obtained by public contributions. Ownership of the solar-augmented heat pump heating and cooling system would ultimately be transferred to the Club's board of trustees and/or the municipality of Scottsdale after installation and completion of performance evaluation of the solar heating and cooling system. Construction of the Scottsdale Girls' Club was to be initiated in late 1974.

The Reference System F design was selected since it permits utilization of currently available subsystems and components. In addition, the Reference System F will act as a precursor to the development of Reference System E, since it incorporates many of the design features of the latter system.

Furthermore, upon completion of the variable pressure ratio heat pump development work, the Reference system F can be retrofitted with this more advanced heat pump concept. This should further increase the cost effectiveness of the Girls' Club building by providing increased annual electrical cost savings. Finally, the southwest area of the United States is one of the most attractive climatic regions for the utilization of solar energy. Hence, it is one of the areas which could lead to the early commercialization of solar heating and cooling systems.

Variable Pressure Ratio Heat Pump

Description: The solar augmented variable pressure ratio heat pump shown in Figures 8.14 and 8.15 will be analyzed, designed, fabricated and tested. Component tests will be performed on the motor, centrifugal compressor, fan coil and sink/source coil, coil fans, and controls.

The performance of all individual components will be determined and the overall performance of the heat pump will be evaluated. Simulated heating and cooling loads (including the effects of solar augmentation) will be used to evaluate the performance of the overall variable pressure ratio heat pump.

The Reference Systems D and E utilize two techniques for increasing overall performance, i.e., solar collector loop augmentation and variable pressure ratio compression. In the heating mode the solar collector loop provides input energy to the hot storage so that the source coil of the heat pump is operating at temperatures greater than ambient.

This results in increasing the heating mode COP from a conventional value of 2 to values potentially as high as 5 to 6 depending upon the climatic region involved. Similarly, in the cooling mode by using an unglazed collector to permit night sky radiation, the cold storage can be reduced to temperature values considerably below those available during daytime ambient conditions. This permits the sink coil (i.e., condenser) of the heat pump to operate at temperatures much lower than could be achieved by an ambient air, fan-cooled condenser.

FIGURE 8.14:   HEAT PUMP—COOLING MODE

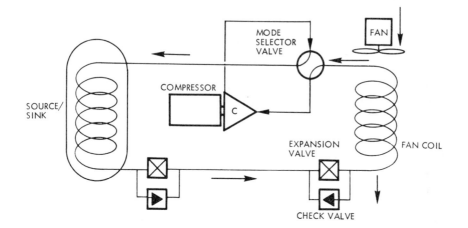

FIGURE 8.15:   HEAT PUMP—HEATING MODE

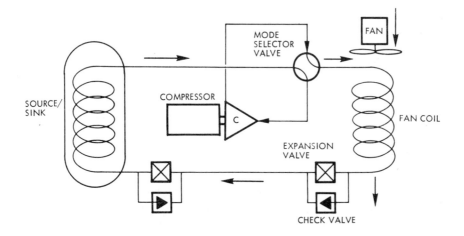

This results in increasing the cooling mode COP from values of 1 to 2 to as high as 4 to 5. Since the amount of electrical energy a heat pump uses is directly related to the COP, a significant saving in electrical energy costs results.

The second technique that can be employed to reduce heat pump electrical energy requirements is to operate with a variable pressure ratio heat pump. Since during many portions of the year the climate is sufficiently moderate, it is possible to operate the heat pump at reduced capacity in order to meet the building loads. Conventional heat pumps or air conditioners utilize a fixed pressure ratio compressor. Thus, at part load the compressors operate at off-design-load conditions which result in reduced compressor efficiency. This results in little or no electrical energy savings despite the fact that the heat pump is operating at reduced capacity. To offset this, it is proposed to develop a variable pressure ratio heat pump.

The system consists of a variable speed centrifugal refrigerant compressor, a fan coil heat exchanger of standard tube-fin construction, a source/sink heat exchanger constructed from copper tubing, a mode selector valve, two expansion valves, and two check valves.

In the cooling mode (Figure 8.14), refrigerant gas from the fan coil heat exchanger, which is acting as an evaporator, is compressed by the centrifugal compressor and delivered through the mode selector valve to the source/sink coil where the gas is condensed. The high pressure liquid then flows through the check valve at the source/sink coil and expands through the expansion valve at the fan coil heat exchanger. The liquid is then evaporated in the fan coil heat exchanger and sent through the mode selector valve to the refrigerant compressor. In this manner, heat is pumped from the fan coil to the source/sink coil.

In the heating mode (Figure 8.15), refrigerant gas from the source/sink heat exchanger, which is acting as an evaporator, is compressed by the centrifugal compressor and delivered through the mode selector valve to the fan coil heat exchanger where the gas is condensed. The high pressure liquid flows through the check valve at the fan coil heat exchanger and expands through the expansion valve at the source/sink coil. The liquid is then evaporated in the source/sink heat exchanger and sent through the mode selector valve to the refrigerant compressor. In this manner, heat is pumped from the source/sink to the fan coil heat exchanger.

Three types of refrigerant compressors have been considered for this application. Two of these are positive displacement type (i.e., piston and screw) compressors. The other is a high speed centrifugal type. Based upon performance and cost tradeoffs the centrifugal compressor was selected.

The centrifugal compressor can provide both variable pressure ratio and good efficiency over a large range of flow rates. This can be accomplished primarily by varying the speed of the compressor. The centrifugal compressor undoubtedly best suits the requirements of the heat pump system. However, a suitable variable speed drive must be developed at a reasonable factory cost to make this solution practical.

For this application, the centrifugal compressor impeller will rotate at approximately 70,000 rpm at its maximum design point and at 50,000 rpm at its minimum pressure ratio requirement. Over this range of speeds the compressor can provide essentially constant efficiency. The problem is to produce a low cost variable speed drive for the impeller.

A 2-pole, 60 Hz motor, will produce a maximum rotational speed of only 3,600 rpm and therefore a speed ratio device of 19.4 to 1 is required. This could be accomplished through a gear box similar to those used for jet engine starters. However, a variable ratio is still required. Motor speed controls of this size would be too costly. A mechanical solution with the potential of meeting a reasonable cost goal is the variable ratio traction drive. This type of drive uses rotating cones which can be positioned to vary the effective gear ratio. The traction drive operates similar to a friction drive except that the elements do not make contact and transfer torque through an oil film.

An electronic approach to the variable speed drive appears to be the most cost competitive at this time. This approach consists of a frequency converter and high speed motor. In this manner a 2-pole motor with a frequency of approximately 1200 Hz will produce the 70,000 rpm required and by varying the frequency over the range of 800 to 1200 Hz the complete operating range can be obtained.

The reason that the frequency converter offers a cost competitive solution, is that the high speed motor can be made extremely small thereby minimizing its cost. The compressor impeller and housing is also very small and can be manufactured at much less cost than the typical piston elements. This leaves a reasonable budget for the frequency converter. Refrigerant R114 was selected to be compatible with the centrifugal compressor requirements and also maintain positive loop pressure over the operating range.

Rationale for Selection: The potential savings in electrical energy by use of a solar augmented heated pump can be of considerable magnitude. By narrowing the overall temperature difference between the sink/source coil and the fan coil of the heat pump a considerable increase in COP can be achieved. In addition by operating the solar collector at modest temperature increases above ambient, high solar collector efficiencies can be obtained.

This latter operating mode can permit the use of unglazed collectors or smaller area glazed collectors. This will result in reduced system costs and enhance the market potential for both Reference Systems D and E. Finally the improved performance achievable with a variable pressure ratio heat pump can also be used with conventional systems thus resulting in a large electrical energy saving for the entire country and opening up a broad new market for retrofitting existing buildings using conventional, low performance heating and cooling systems.

West Los Angeles City Hall

Description: The Reference System B (see Figure 8.4) configuration will be analyzed, designed, fabricated, tested, and then installed and integrated into the existing West Los Angeles City Hall Building. The solar collector will be based upon a TRW unglazed, low cost configuration but will be adaptable to the installation of a single glaze design if required. The circulating fluid will be water and a hot water energy storage tank will be specified and procured for use with this system.

The auxiliary hot water heater, pumps, heating coils and controls will all be specified and obtained from commercial sources. Performance tests to evaluate solar collector efficiency for both the unglazed and glazed designs will be conducted. In addition, percent solar energy utilization will be determined for both these configurations.

The West Los Angeles Municipal Building is approximately 72 feet wide by 257 feet long and 2 stories above grade. The building is heated and cooled by air handling equipment located in a partial basement and a penthouse 32' x 112' on the roof. The balance of the roof area above the second floor and the roof over the penthouse are available for locating the solar collectors. The mechanical equipment room is adequate for location of hot water storage tanks, while the four multizone two deck air handling units can be readily modified to incorporate heating (preheat) coils (heat exchangers) which would be part of SES.

The building is ideally suited for retrofitting Reference System B with glazed or unglazed collectors and offers a high potential for demonstrating Reference System E where a heat pump would be employed. The peak heating load is about 750,000 Btu/hr.

Rationale for Selection: The West Los Angeles City Hall will be made available for retrofit with a solar hot water and space heating system based upon the Reference System B. This building meets several important criteria for its acceptability for a Phase 1 POCE. It is a government owned building and transfer of ownership of the solar hot water and heating system to the West Los Angeles municipality (after installation and completion of performance evaluation) is anticipated. The building is located in a climatic region that is ideally suited for solar energy system use. In addition, the region also possesses a high market capture potential based upon both new construction and demographic considerations.

The implementation of a Reference System B type configuration will provide important performance data for both unglazed and single glazed collectors as well as energy storage subsystem characteristics, which will be used to establish a minimum cost design of Reference System D for use with the future planned Los Angeles National Bureau of Standards building addition.

These data, together with the data obtained from the variable pressure ratio heat

pump program, should prove invaluable in designing a cost effective Reference
System D for implementation in Phase 2, with minimum business and development risks. The emphasis in arriving at this design configuration will be to minimize the initial cost of the system and to maximize the savings in electrical
energy consumption.

FUTURE DEVELOPMENT PLAN

From among the fourteen cities selected during the study, six were chosen for
nation-wide implementation of the system designs developed. The cities are
Los Angeles, Phoenix, Boston, New York, Dallas, and Denver. The six cities
were selected on the basis of market capture potential, which in turn was based
on climatic conditions, conventional energy types and uses, projections of future
building starts and conventional energy costs. Four of the five original building
types were selected and a total of nine buildings are recommended for each city
as follows: multifamily low-rise apartment buildings (2), single family residences
(5), schools (1), and commercial buildings (1).

The residential units were emphasized since this segment must demonstrate a
high market capture potential if significant total energy savings are to be realized.
The multiple family residence is already one of the better markets, but the single family residence has one of the lowest capture rates of the five building types.
This is somewhat of an artificial situation since a major reason for the difference
in capture potential between the two types of residences is the preferred electric
utility rate given to single family units. Schools have been selected because they,
too, have high capture potential.

Although the total market is relatively small because of the number of schools
and anticipated new starts is small, the exposure value in terms of stimulating
public awareness and acceptance of solar energy systems is high. Office buildings
have been included even though the market capture potential is expected to be
less than 10%. However, most government buildings (including federal, state,
and local) fall into this category and are prime candidates for solar energy systems, both with regard to life-cycle cost considerations and public exposure
value.

The design, construction, and operation of these 54 buildings should provide
the performance data and public exposure required to successfully initiate widespread utilization of solar energy systems for the heating and cooling of buildings.

UTILIZATION PLANNING

In order to stimulate the introduction of SESs on a significant scale, the following concurrent activities should be carried out. The government must play a
strong role in SES industry development. Specific focus should be on R&D

funding oriented towards reducing SES costs and on funding for demonstration projects. Public opinion surveys clearly showed that consumers in cities with solar demonstration projects are significantly more aware of the concept of solar heating and cooling and are more likely to accept their utilization. Thus, demonstration projects should be planned for buildings that have high public exposure, such as schools, government buildings, etc.

Additional government participation must focus on implementing building code modifications requiring increased energy conservation, and performance rather than prescriptive building code criteria. Government assistance in establishing standards and performance criteria for SES would reduce insurance industry opposition. Government can also play a large role in minimizing financial institution resistance. The Federal Home Loan Bank Board can use its influence to encourage member financial institutions to grant mortgages on SES homes. Furthermore, the government can insure the SES incremental mortgage amount, thus diminishing financial institution reluctance to finance this increment.

A major impact on both consumers and developers could be generated through the government requiring SES installations on government-funded projects, especially housing projects. SES installation on government housing would generate consumer awareness, but more importantly, would also demonstrate the direct practicality of SES for residences.

A further role the government can play is in offering incentives to consumers and/or to SES manufacturers. Possible incentives include tax credits, low-interest financing, and tax deductions. It will also be important to assure that state and local property taxes on the incremental costs of SESs are not applied in a manner that produces a negative incentive.

An important role must also be assumed by industry organizations such as NAHB, ASME, and ASHRAE. These organizations have a large direct effect on developer and other decision maker acceptance of SES.

The housing market can be approached most advantageously by eliciting SES usage by large tract developers. Most new construction is done on a volume basis, and necessarily provides the largest markets. Equally important, small builders tend to follow the lead of large builders, but will generally be more conservative initially. However, the custom home market should not be overlooked because the buyer is generally less first-cost sensitive. Also SES is more financially feasible for luxury condominium apartments than for investor-owned and operated apartments.

Another important participant in any implementation plan is the utility industry, particularly the gas companies. Since the supply of natural gas is limited and some regions now have moratoriums on new hookups, the use of solar energy is an ideal approach for permitting the expansion of the customer base without increasing the consumption of natural gas.

The involvement of professional builder/owners of industrial and commercial buildings could be very beneficial for the implementation of solar energy systems. Since this group is very sensitive to life-cycle costs, they could be expected to be among the most likely early users of these systems.

CONCLUSIONS

Solar heating and cooling of buildings provides a potential market approaching a billion dollars per year by the year 2000, even under the conservative assumptions of the capture analysis and measured in terms of 1974 dollars.

Solar-energy-augmented systems can be designed to permit a 1% reduction in total national conventional energy consumption by the year 2000. However, this requires the development of construction-integrated solar energy collection systems (to lower the incremental capital costs) in conjunction with variable pressure ratio heat pumps (to provide improved coefficients of performance and thereby reduce conventional fuel consumption and operating costs). The POCE program recommended above is directed towards determining the system and subsystem designs most likely to meet the objective of attaining significant conventional energy savings in the heating and cooling of buildings. The Future Development Plan is oriented towards implementing these system designs on a nationwide scale.

Colorado State University Integrated System

The first integrated system providing heating and cooling to a building by use of solar energy has been designed and installed in a residential type building at Colorado State University. The following material describes this integrated system as reported in NSF-RA-N-74-104, *Design and Construction of a Residential Solar Heating and Cooling System,* by G.O.G. Löf, D.S. Ward, J.C. Ward and C.C. Smith of the Solar Energy Applications Laboratory, Colorado State University and the Solar Energy Laboratory, University of Wisconsin. This report was prepared for the National Science Foundation and issued in August 1974.

PROJECT OBJECTIVES AND BUILDING DESIGN

The primary objectives of the project were to:

1) Establish the practicality of space cooling with solar energy;

2) Design an effective and economical system for residential heating and cooling and service hot water heating with solar energy,

3) Construct a fully automated and instrumented solar heating and cooling system in a new residential building;

4) Appraise the performance of the complete heating and cooling system and each of its principal components,

5) Determine which of several operating modes minimizes auxiliary energy requirements,

6) Appraise the utility of a mathematical model in design and prediction of performance by comparing actual and predicted results, and

7) Modify system design and operation to minimize total annual cost of heating and cooling.

232

To accomplish these objectives, the following principles and criteria were established:

1) Use of an optimized system of predictable satisfactory performance with currently known technology,

2) Selection of a building design which has "typical" rather than "extreme" or "unusual" appearance, size, form, materials, energy requirements, insulation, fenestration, and all other characteristics of a modern single family house,

3) Provision of a comfort level in the building fully equal to that obtainable with the best conventional heating and cooling systems,

4) Design of building so that the solar collectors and other components can be replaced with completely different units for testing after the present system has been fully evaluated,

5) Design of collector suitable for mounting on the roof surface but also suitable for modifying and roof replacement,

6) Provision for about three-fourths of the annual heating load and three-fourths of the annual cooling load with solar energy,

7) Provision of sufficient flexibility and redundancy in the heating and cooling system to permit comparison of several variations in design and operation method,

8) Use of commercially available controls in a fully automatic system requiring no human intervention other than setting a wall thermostat for the desired house temperature,

9) Establishment of reliability, durability, and safety at higher priority than least cost in this experimental installation,

10) Selection of systems providing lowest cost without jeopardizing performance or durability, and

11) Emphasis on designs and materials which could provide low cost solar energy with high volume manufacture.

The foregoing objectives and principles were met by designing a solar heating and cooling system in a residence type building at a site on the Foothills Campus of Colorado State University in Fort Collins, Colorado. The house is situated at 40.6° N and 105.1° W, at an altitude of 1,585 meters (5,200 feet) above sea level. The mean annual heating degree days are approximately the same as the 3,322°C-days (5,980°F-days) total in Denver. The design temperature (for the Denver area) is -23°C (-10°F) (winter) and 35°C (95°F) dry bulb, 18°C (64°F) wet bulb (summer).

A suitable building for the experimental solar heating and cooling system was provided by constructing a modern three bedroom frame residence with a living area of 140 square meters (1,500 square feet) and a full, heated basement. The design heating load was computed to be 16.1 kilowatts at -23°C (55,000 Btu/hr at -10°F, corresponding to 17,600 Btu/Deg F Day). The design cooling load is approximately 10.5 kilowatts (3 tons or 36,000 Btu/hr). The insulation was typical, with 8.9 cm (3½ inches) fiber glass in the walls and 14 cm (5½ inches) of fiber glass in the ceiling. The building design included several energy conservation and solar design factors which are as follows.

Architecture: location of structure for maximum solar utilization, roof angle optimized for solar collector heating and cooling, favorable solar configuration and plan, basement principally below grade, garage placed as wind and storm buffer, air locked entries, designed access to solar collector collection system and instrumentation, weathertight construction and ease of maintenance.

Fenestration: solar shading control over glass areas, weathertight sash, all glass areas double glazed with insulating glass and limitation on north and west window areas.

Insulation: roof insulation (U = 0.03), wall insulation (U = 0.07), vapor barrier provided and weatherstripping of openings.

Convection: external convection controlled by building location, configuration, topography and landscaping and internal convection by stairways, external openings, spatial volumes and variation in ceiling heights.

Surfaces: exterior surface to help retain "thin air film" around building, dark surface for collectors and internal light color surfaces for maximum light reflection.

Passive Solar: south porch decks temper incoming air, garage entrance exposed to sun and solar radiation enters glass in winter on south elevation but not in summer.

An exterior view of the structure is shown in Figure 9.1.

FIGURE 9.1: COLORADO STATE UNIVERSITY SOLAR HOUSE

DESIGN OF HEATING AND COOLING SYSTEM

General Description

The solar heating and cooling system is shown in simplified form in Figure 9.2. Conventional components are the lithium bromide absorption cooling unit, hot water boiler, air heater coil, hot water heater and associated piping, ducts, and pumps. The solar components consist of a solar collector and pump, thermal storage with heat exchanger and hot water preheat tank, and an automatic valve.

The actual installation at Colorado State University is more complex and provides considerable flexibility in operating modes. Figure 9.3 is a cross-section schematic diagram of the installation which shows all of the components except the collector and associated piping, the control sensors, and the air distribution system. The primary modes of solar heat collection are storing heat from the solar collector via a heat exchanger and storing heat directly from the solar collector. A third mode of solar collection is by supplying heated fluid directly from the solar collector to the cooling unit. Energy to the heating or cooling unit is provided either by use of hot water from storage, if the temperature is adequate, or from the auxiliary boiler as necessary.

An alternate heating mode utilizes whatever heat is in storage, even at temperatures as low as 27°C (80°F), with the auxiliary boiler supplying hot water to an auxiliary air heating coil. In this mode the auxiliary boiler acts as a temperature booster for the solar system. In all other modes of operation, either the auxiliary or solar is used, never the two together.

The house hot water system utilizes solar storage for preheating service hot water. Water from a cold water main enters the preheat tank, to which heat is supplied from solar storage by circulation through a heat exchanger. On demand, the preheated water then flows to a conventional gas hot water heater, which maintains the required temperatures.

Because of the possibility of freeze damage by circulating water through the collector, normal operation provides solar heat collection in a 60% solution of ethylene glycol (commercial automotive antifreeze) in water. The cost of several hundred gallons of glycol in the main storage system is prohibitive, so a heat exchanger is employed for transfer of heat from the small volume of collector fluid (about 28 gallons) to a large volume of water comprising the thermal storage.

The use of a liquid medium for solar collection and water for thermal storage is dictated primarily by requirements of the cooling unit. Additional considerations include the ready availability, experience with its use, high heat capacity, and low cost. The use of another storage liquid would require determination of its compatibility with the various components of a heating and cooling system (pumps, piping, heat exchangers). The limited availability of components specifically designed for use with another fluid is an additional factor.

FIGURE 9.2: SIMPLIFIED SOLAR HEATING AND COOLING SYSTEM

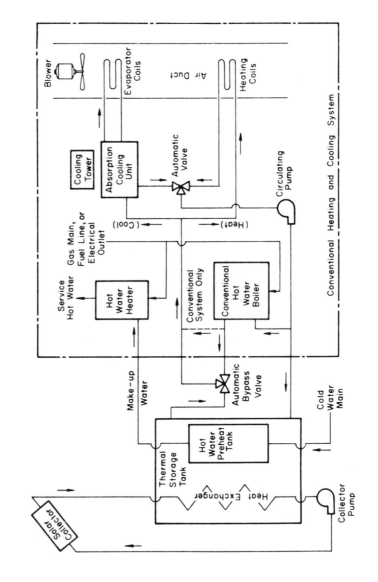

FIGURE 9.3: SOLAR HEATING AND COOLING SYSTEM

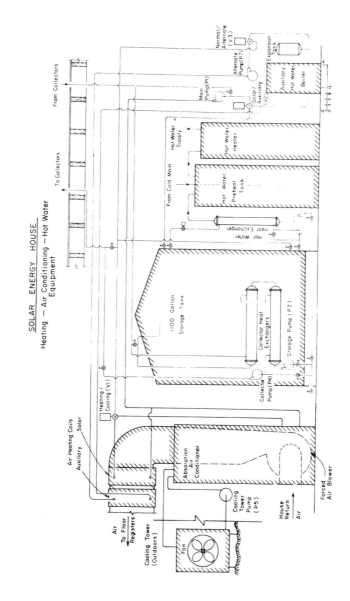

SOLAR ENERGY HOUSE
Heating — Air Conditioning — Hot Water
Equipment

Figure 9.4 details the interface between the collector panels and the rest of the solar system. The collector absorber panels are made of aluminum and must therefore be isolated from the copper piping in the solar system. This is accomplished with a filter, an aluminum getter (sacrificial), and rubber hose connections between the copper and aluminum piping.

A 30-gallon surge tank is installed on the outlet side of the collector. This allows for boiling in the collector (and the subsequent pressure build-up). For example, if a power failure occurred during the day and the collector pump shut down, the collector would boil the liquid out of the system. The surge tank allows for this. Then when the power returns, the surge tank will refill the system automatically.

The storage container is a light gauge vertical galvanized steel cylinder 1.83 meters (6 feet) high and 1.68 meters (5½ feet) in diameter. It holds 4,275 liters (1,131 gallons), a nominal 61 liters per square meter of collector (1.5 gallons per square foot). Surrounding the tank are two layers of bonded glass fiber double-faced batt insulation.

The size of the thermal storage container is primarily an economic decision. It is far too costly to meet the full heating and cooling load with solar energy, so the essential requirement of storage is to accumulate solar heat in the daytime for use at night. Overnight storage is clearly advantageous since this comparatively low cost unit permits nearly continuous solar use in normal, sunny weather. The collector can thus be sized for carrying night loads as well as daytime requirements.

Storage for a full 36-hour sunless period (two nights and one intervening day) involves a large additional cost for comparatively small benefit and would seldom be cost effective. Furthermore, the frequency of two or more successive cloudy days is not sufficient to justify storage of more than one day of midwinter heat demand. If more storage were provided, additional collector area would also be needed to heat it during the preceding sunny period. The added collector surface would have an even poorer utilization factor and is therefore not economically viable.

The storage tank must be operated at atmospheric pressure because of the light gauge material used. Although this provides a capital cost saving, operating problems must be dealt with. In the cooling mode, water must be pumped to the generator at 170° to 203°F (local boiling point). Operating in this range near boiling, only slight deviations in pressure can produce vapor lock in the pumps. Such conditions can interfere with effective pumping unless suitable hydrostatic head is provided.

The problem is most noticeable in the auxiliary boiler, where the head loss across the boiler and near boiling temperatures can produce a vapor lock at the pump intake. Pumping from solar storage is less critical because the head on the suction side of the pump is about 5 feet. But for designs using storage below ground, this factor becomes of serious concern.

FIGURE 9.4: PIPING AT COLLECTOR INLET AND OUTLET IN ATTIC

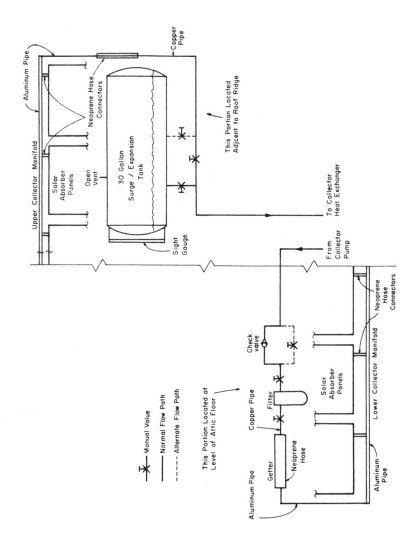

The use of a nonpressure container requires a vent to allow for boiling conditions within the storage container. In the Colorado State University design, the tank is vented outside the building to avoid adding to the cooling load. The venting also requires the introduction of make-up water. Water from the house water system is supplied via a float valve in the storage tank.

Very little stratification is possible in the storage tank. Flow rates through the collector heat exchanger or directly through the collector loop are too high for maintenance of stagnant water layers in storage. The benefits of stratification were determined to be insufficient to justify the installation of baffles or other devices to retard mixing in the tank.

A standard hot water boiler as the auxiliary heat source has the advantages of simple interfacing with the rest of the solar system and ready commercial availability. The size of the unit is determined in a normal fashion by the maximum heating/cooling demands of the building. The temperature of the hot water in the boiler will be maintained at approximately 190°F (90°C) for direct use in the absorption cooling unit and heating coil.

The manner in which auxiliary heat is supplied affects over-all system performance and cost. The auxiliary may be used to boost the temperature of the hot water coming from the heat storage subsystem, if its temperature does not meet heating and cooling requirements, or the auxiliary may be used exclusively when storage temperatures are too low to be useful.

A third method involves using stored solar heat and auxiliary heat simultaneously, in independent water-to-air heat exchangers.

A fault in the first method can be shown by examining the system when used for cooling. If the storage temperature is 170°F, the auxiliary could be used to raise the temperature of the water to 190°F, a 20°F rise. But the water temperature drop through the generator is only 10 degrees or less. Hence, storage water at 170°F would be heated to 190°F and returned to storage at 180°F, thereby using some of the storage heat capacity for fuel energy. Collector efficiency and useful solar storage could both be reduced by this procedure.

The same situation would be encountered in the heating mode, so this manner of auxiliary use will not be employed. Normal operations will involve use of auxiliary heat directly to the loads, not as a storage temperature booster. Specifications of the equipment used for the heating and cooling system are described in the section which follows.

Specifications of Equipment

Collector: The collector consists of sixteen 0.9 meter by 4.9 meters (3 feet by 16 feet) panels. The total collector area is 71.3 square meters (768 square feet). The total absorber surface is 66.9 square meters (750 square feet). The total exposed glass area is 65.7 square meters (707 square feet). Manifold pipes

5 centimeters (2 inches) at the top and bottom of the collector area run the length of the collector (14.6 meters, 48 feet).

Circulating Pumps: The pumps are all centrifugal type with direct coupled motors. They are all 1,750 rpm constant speed. Centrifugal pumps provide two advantages over the positive displacement style pump. Centrifugal pumps offer a safety feature in that they will pump only a small amount above rated pressure if the fluid loop should be blocked. Thus such an occurrence would neither damage the pump nor burst a fluid line. A second advantage, particularly on the collector loop, is the increase flow rate as the temperature of the fluid increases. This is due to viscosity changes of the fluid and improves the collector efficiency at high temperatures.

The hot water preheat and cooling tower pumps were stock manufactured pumps which met the requirements without any modifications. The other pumps in the system were adapted to specific pressure (head) versus flow rate requirements. This was accomplished by using pumps slightly larger than needed and reducing the diameter (trimming) of the impellers to obtain the exact specifications. This was done by the manufacturer at no cost and is a common procedure. The following are specifications for the liquid pumps.

Pump	Flow (GPM)	Head (ft H_2O)	Horsepower	Manufacturer
Main System (P1)	11	24	1/5	Bell & Gossett
Storage (P3)	25	7	1/6	Bell & Gossett
Collector (P4)	16	34	1/2	Bell & Gossett
Cooling Tower (P5)	10	35	1/3	Burks
Alternate (P7)	5	8	1/12	Bell & Gossett
Hot Water (P2, P6)	2	2	1/40 ea.	Teel

Heat Exchangers: For liquid to liquid heat exchange without mixing of the liquids, shell-and-tube type heat exchangers were selected. These units are marketed widely and were thus easily obtained. The only uncommon requirement for these heat exchangers was for a low temperature difference between the two liquids. This requirement was met by using single-pass counterflow design. This is illustrated in Figure 9.5.

It can be seen from the temperature profiles in the exchanger that the single-pass counterflow arrangement allows the temperature difference between fluids to be nearly constant along the exchanger. This feature results in a smaller temperature loss across the exchangers. The disadvantage of the single-pass counterflow heat exchanger is its physical dimensions, involving high length and low diameter. The collector heat exchanger is made up of two units in series because a single unit of sufficient length was unavailable. A second disadvantage is the high flow rate required through the tubes of the exchanger.

FIGURE 9.5: SINGLE-PASS COUNTER FLOW HEAT EXCHANGER

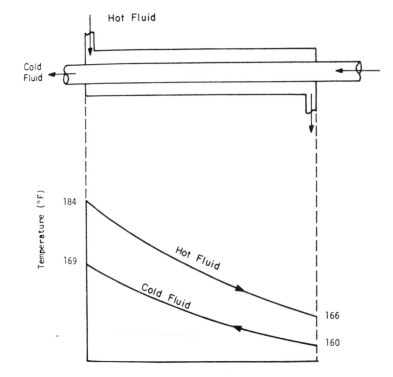

The single-pass design results in more tubes in parallel. This means that a higher pumping rate is needed to develop turbulent flow in the tubes. It may be noted that the tube liquid in the collector heat exchanger is the storage tank water and is pumped at 25 gpm. This flow is obtainable with modest pump power because resistance in that loop is low.

The high flow rate does, however, nearly eliminate temperature stratification in the storage tank. Highly stratified storage temperatures would thus come at the cost of a larger exchanger or a higher temperature loss. Specifications for the liquid to liquid heat exchangers are given in the table at the top of the following page. The heat exchangers were supplied by Young Radiator Company.

Heat Exchanger	Shell			Tube			Design Heat Rate (btu/Hr)	UA Btu/Hr - °F
	Flow (GPM)	Temp. Drop (°F)	Pressure Drop (psi)	Flow (GPM)	Temp. Drop (°F)	Pressure Drop (psi)		
Collector	16	18.0	9.0	25	9.2	.2	115,500	12,500
Hot Water	2	19.6	.25	2	19.6	.25	20,000	1,000

Air Duct Coils: Solar heating requires a larger air duct coil capacity than does auxiliary boiler heating. This is due to the lower temperatures from solar storage (down to 75°F in the alternate mode) compared with the boiler water at 190°F. The air duct coil size required for solar heating was thus selected and the same size coils were obtained for both solar and auxiliary because of easier installation and smaller air pressure loss with the identical physically sized coils. The coils consist of copper tubes with aluminum fins housed in a 20 by 22 inch duct section. The air duct coils are stock manufactured units (Model DCH 36-90) supplied by Arkla Industries, Incorporated. The specifications are as follows.

Water Flow (GPM) Temp. Drop (°F)	Air Flow (CFM) Temp. Rise (°F)	Design Heat Rate (Btu/Hr)	Air Pressure Loss (in H_2O)
10 13.7	1200 67.5	70,000	.14 (each)

Hot Water Heater and Preheat Tank: The hot water heater acts as a gas fired back-up for service hot water. It is a 40-gallon glass lined tank with a 35-gallon-per-hour recovery rate at 100°F temperature rise. The output is rated as 42,000 Btu/hr. The hot water preheat tank is a standard 80-gallon glass lined electric hot water heater with the heating element and control nonoperational. This proved to be the least expensive and most readily available insulated pressure pressure tank.

Auxiliary Hot Water Boiler: The hot water boiler is a stock manufactured unit for domestic hot water heating systems. It was selected to provide the full heating and cooling requirements for the building as an energy back-up for periods of cloudy weather coupled with high energy demand. The boiler is gas fired with a rated input of 90,000 Btu/hr and output of 72,000 Btu/hr, supplied by the Crane Company. At the 5,000-foot altitude of Fort Collins, corresponding ratings are 72,000 Btu/hr and 57,600 Btu/hr, respectively.

Solar Storage Tank: The storage tank was fabricated from 16 gauge (0.0598 inch) galvanized sheet steel. Seams and pipe connections to the tank are arc

welded. A two foot diameter manhole in the top of the tank allows complete
access to the inside. Each piping connection on the tank is provided with a
shut-off valve and a neoprene hose connection to the copper piping. The tank
is electrically isolated from all other plumbing components to prevent electro-
lytic corrosion of the tank. Because iron and zinc are more active than copper
on the electromotive scale, storage tank metal would otherwise be removed
and deposited on the copper. The specifications of the storage tank are as
follows.

Diameter	5.5 feet	(1.67 meters)
Height	6.0 feet	(1.82 meters)
Height of top cone section	0.9 feet	(0.27 meter)
Volume	1,100 U.S. gallons	(4,164 liters)
Weight empty	470 pounds	(213 kilograms)
Weight filled	9,644 pounds	(4,374 kilograms)

Collector Filter: The collector fluid is filtered by an in-line filter. It is Model
CA manufactured by the Ferro Corporation of East Rochester, New York.

Collector Getter: The collector ion getter was fabricated at CSU by use of
aluminum screen wire rolled into a neoprene hose 2½ inches in diameter and
1½ feet long. The getter follows the filter and is the last element before the
fluid enters the collector (see Figure 9.4).

SYSTEM OPERATION AND CONTROL

Due to research requirements of this project, a high degree of versatility has
been designed into the mechanical system. By such design, various modes of
operation will be appraised and the one which provides maximum advantage
can be selected. The primary objective of mode selection is to determine how
solar heat can best be utilized and the total costs of fuel and capital minimized.

The data acquisition system is designed to accommodate each mode with complete
measurement. Differences in performance between the different modes can thus
be determined. The automatic control system is also designed to accommodate
the mode selection. After a mode is manually selected on the control panel, the
system is automatically controlled in that mode. The nature of the several modes
and their probable advantages and disadvantages are outlined below.

Mode 1

Mode 1 is characterized by the use of a heat exchanger which separates the
collector fluid from the storage tank fluid. Figure 9.6 illustrates with bold
lines the fluid circuits used in Mode 1. The collector heat exchangers are commercially available shell-and-tube type. In this application the collector fluid
is pumped through the shells or outer passages while the storage fluid is pumped

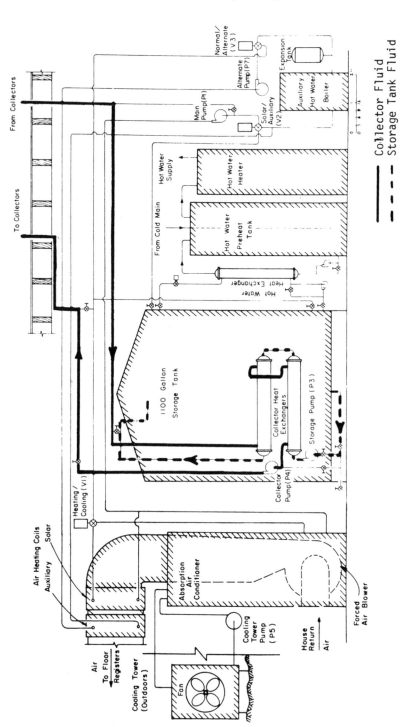

FIGURE 9.6: MODE 1—SOLAR COLLECTION USING HEAT EXCHANGER

through the tubes. Two single-pass exchangers were chosen to provide a practical minimum difference in temperature between solar collector and storage tank.

The two main advantages of Mode 1 are the avoidance of antifreeze in the storage system and the use of nonpressurized (vented to the atmosphere) storage tank. The only acceptable antifreeze material for this service is an ethylene glycol base fluid with corrosion inhibitors such as used in automobile cooling systems. With the 60% antifreeze mixture required for adequate antifreeze and boil-out protection, a total of 680 gallons would be needed. Replacement of the fluid on a semiannual basis, as suggested by some manufacturers, or even at less frequent intervals, would impose an unacceptable expense to the householder.

Provision of the heat exchanger in Mode 1 permits use of only 15 gallons of antifreeze in the collector loop. The storage tank then needs only water and a corrosion inhibiting additive. The need for antifreeze could be eliminated, however, if the collector fluid could be made to drain out of the collectors automatically during periods of freeze danger to the absorber plates. Although this capability was designed into the Colorado State University solar house collector panels, the reliability of complete drain-down has not yet been determined.

The second major advantage of Mode 1 is a pressurized collector circuit and a nonpressurized storage tank, with the heat exchanger acting as the pressure barrier. There are two reasons for preferring a pressurized collector loop. Under pressure, the boiling temperature of the collector fluid is elevated. This allows higher operating temperatures in the collector, which can be of particular advantage for summer air conditioning. Also with the heat exchanger preventing pressure in the storage tank, the collector loop can be completely fluid filled. Consequently, there is no gravity head loss for the pump to overcome, only frictional head loss.

To "spill" the fluid from the top of the collectors to an open storage tank (a height of 25 feet) would in this case require doubling the collector pumping power. A tank capable of withstanding this hydrostatic head and large enough to accommodate 1,100 gallons plus a thermal expansion air volume would cost three to four times as much as a nonpressurized tank.

A third advantage of Mode 1 concerns corrosion protection of the collector absorber plates. The need for continuous filtration and deionization of the collector fluid is a major consideration. Mode 1 requires only the 28 gallons filling the collector loop to receive such treatment rather than the additional 1,100 gallons of fluid in storage.

Mode 2

Mode 2 provides collection of solar heat directly in the storage tank. It does

FIGURE 9.7: MODE 2–SOLAR COLLECTION WITHOUT HEAT EXCHANGER

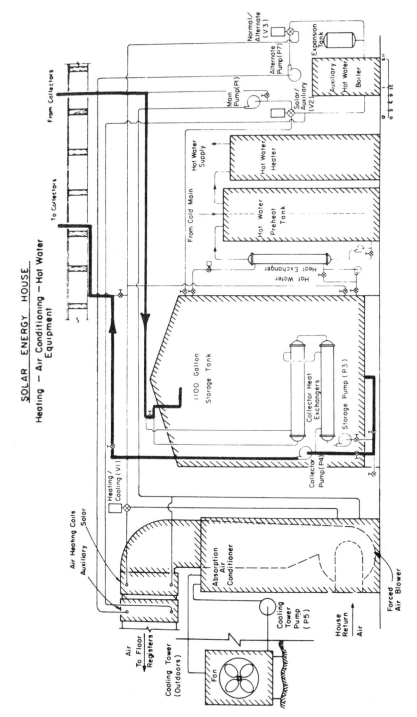

SOLAR ENERGY HOUSE
Heating — Air Conditioning — Hot Water
Equipment

not utilize the collector heat exchangers or the storage pump required in Mode 1. Figure 9.7 illustrates the fluid circuit for collection of solar heat in Mode 2.

Two advantages are recognized with Mode 2. There is the saving of capital and maintenance costs for the collector heat exchangers and the storage pump. A second advantage of Mode 2 is the elimination of the temperature drop across the collector heat exchangers (ranging from $0°$ to $6°C$). The net result is an improvement in system efficiency. The University of Wisconsin computer simulation has been run both with and without the collector heat exchangers (Modes 1 and 2). The results of these runs show about 1% more of the total heat load supplied by solar heat for the entire year using Mode 2 rather than Mode 1.

Mode 3

Mode 3 is a cooling design which does not utilize heat storage but supplies solar heated fluid directly to the generator of the air conditioner. Figure 9.8 is a diagram of this mode.

The advantages of Mode 3 are higher water temperatures available to the generator of the air conditioner and the avoidance of some heat loss into the house from the storage tank. Heat loss from the storage tank in summer is significant because it has a dual effect on cooling operation. Not only is less heat available to the generator but heat lost into the house must be made up by the air conditioner to maintain the desired house temperature. The storage tank is well insulated, but some loss is inevitable. Measurements and calculations show that about 30,000 Btu are lost to the house from the storage tank per day. At the rated 3-ton (36,000 Btu/hr) cooling output of the air conditioner, this is equivalent to about an hour of air conditioner operation. Air conditioning directly from the collectors would mean a somewhat lower storage operating temperature and a smaller heat loss to the building.

A disadvantage of this mode is the lack of stored energy for cooling in the evening. This is not a problem at the Colorado State University solar house location, however, because the need for cooling after sunset diminishes rapidly.

The avoidance of storage tank mixing means that Mode 3 provides higher temperature to the air conditioner generator. This increases cooling output. A problem foreseen with Mode 3 is poor cooling control due to solar input direct to the air conditioner whenever available. Cool storage in the house interior may, however, provide some stabilizing influence. Mode 3 is not practical for winter heating because the timing of heating demand and solar supply requires substantial storage for adequate comfort.

Alternate Mode

In the heating season, there will often be a condition in which the storage tank temperature is considerably above house temperature, but is not high enough to

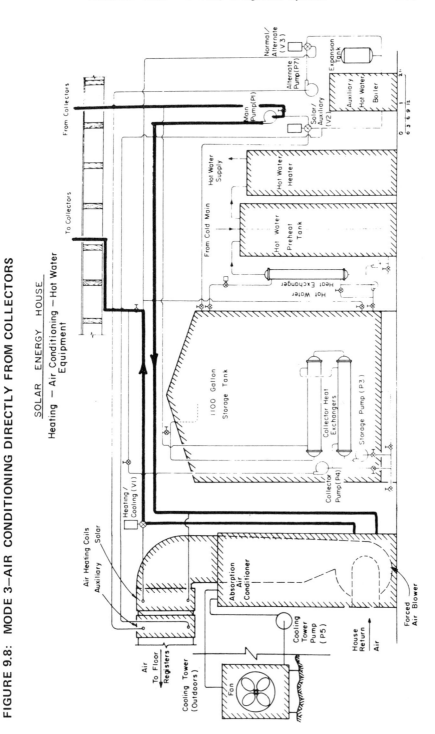

FIGURE 9.8: MODE 3—AIR CONDITIONING DIRECTLY FROM COLLECTORS

SOLAR ENERGY HOUSE

Heating — Air Conditioning — Hot Water Equipment

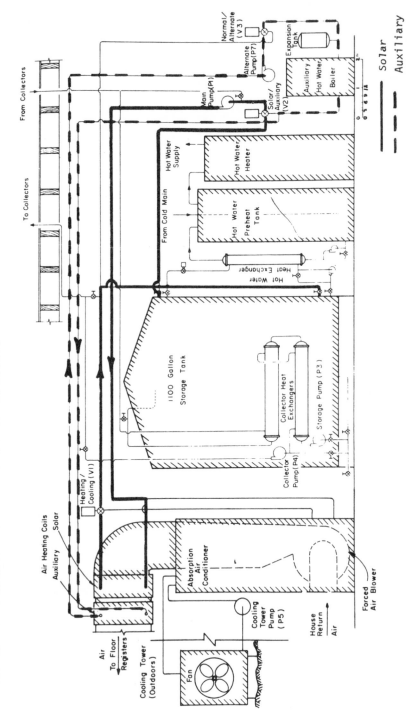

FIGURE 9.9: SOLAR HEATING WITH AUXILIARY BOOSTING

carry the entire heating load (maintain the desired house temperature setting). It is particularly desirable to use this heat at moderately low temperature because it is acquired at high collector efficiency.

The liquid-to-air heating design allows for separate solar storage and auxiliary boiler loops to supply heat in the air duct. This arrangement is illustrated in Figure 9.9. The solar air heating coil is placed ahead of the auxiliary air heating coil in the direction of air flow. The solar coil thus preheats the air while the auxiliary coil boosts the air temperature to that required to maintain the heating load. The net result is the use of more solar and less auxiliary heating. The disadvantage of the alternate mode lies with the cost of an additional air heating coil and the alternate pump.

Automatic Controls

One of the objectives in the design of the Colorado State University Solar House is a fully automatic control system. The occupant of such a house would need only to set the desired house temperature on a wall thermostat and to set the service hot water temperature; no additional adjustment would be required. A corollary objective is a control system providing solar heating, or cooling, and hot water at comfort levels fully comparable with conventional fuel and electric sources.

Extraction of heat from the solar collectors and distribution of heat to the heating, cooling, and hot water loads are automatically controlled in response to preset temperatures or temperature differentials. This is accomplished by feedback control loops consisting of sensors, thermostats, controlled elements and the mechanical system itself.

Figure 9.10 illustrates a typical feedback control loop. The sensors are vapor expansion bulb and capillary aquastats for liquid temperatures, electrical resistance (thermistor) elements for collector absorber plate temperature, and a bimetal coil for air temperature measurement in the house. Thermostats are used to compare the sensor outputs with the corresponding temperature or differential temperature settings. When the sensor output is sufficiently different from the setting (or is outside of the deadband), a control signal at 24 volts is transmitted from the thermostat to a power relay. The relay opens or closes the 110 volt power line to a controlled element which is a pump, motor, or an automatic electric valve positioner, or an air fan motor (see Figure 9.11 for control sensors).

The mechanical system is the final link in the feedback control loop. It is affected by controlled elements which regulate flow to the various components. Operation of the solar air heating coil, for example, depends on the conditions of the main system pump **P1** and the heating/cooling automatic valve **V1**. The mechanical system produces an output measured as temperature by the control sensor, thus completing the feedback control loop. Through operation of the air coil, for example, house air temperature change is sensed by the wall thermostat. All control action is of the on/off type with adjustable set points and dead bands. The use of modulating or proportional control was avoided due to the relatively high cost of the controlled elements requiring feedback servo mechanisms.

FIGURE 9.10: BLOCK DIAGRAM–CONTROL LOOP

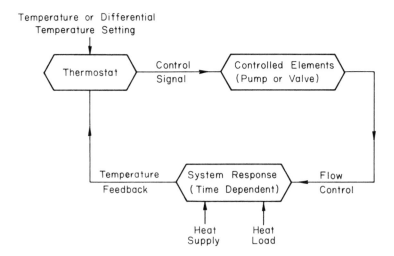

The control system is more complex than would be required for a single mode system. The multimode design was chosen to provide versatility of operation. A mode is selected manually and the control system automatically assumes control in that mode. Honeywell, Incorporated has provided the equipment and engineering design to accomplish the solar house control functions.

COLLECTOR DESIGN

Because of the experimental nature of the installation, and the need to replace the collector with new designs at a later date, a watertight subroof was made an integral part of the house design. This requirement imposed some important restrictions on the possible collector configurations, e.g., the installation of a prefabricated modular unit becomes more difficult when the back side of the collector is unavailable for making plumbing connections. It was also desired to have a neat appearing watertight solar collector capable of acting as a roof. These requirements further limited the possibilities of utilizing modular construction. The design finally developed is shown in Figure 9.12.

The collector is composed of aluminum structural supports, two sheets of B-quality double strength (one-eighth inch) window glass, an aluminum absorber plate with internal tubes (Roll-Bond), and insulation beneath. The cover glasses

FIGURE 9.11: LOCATION OF CONTROL SENSORS

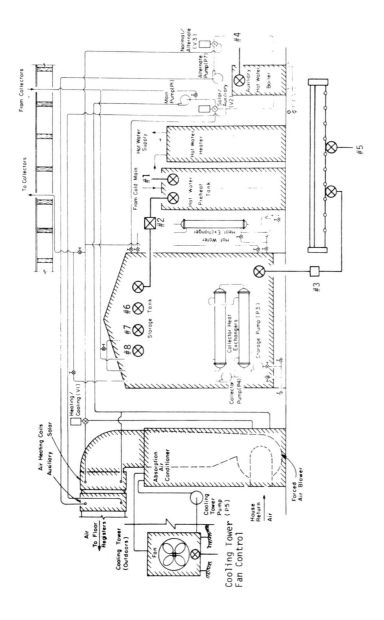

are attached to the aluminum structure by a butyl tape and an aluminum cap strip. Approximately one-third of the glass covers have undergone an antireflection treatment (RCA Magicote process), courtesy of Honeywell, Inc. The aluminum panel has a tube pattern of parallel flow in multiple straight tubes between internal sloping manifolds. A flatback Nextel acrylic coating was baked on the aluminum surface at a temperature of 200°C (400°F). The insulation consists of 2.54 centimeters (1 inch) unbonded glass fiber mat on a 3.81 centimeters (1.5 inches) Fesco-Foam composite insulation.

FIGURE 9.12: SOLAR COLLECTOR CROSS SECTION

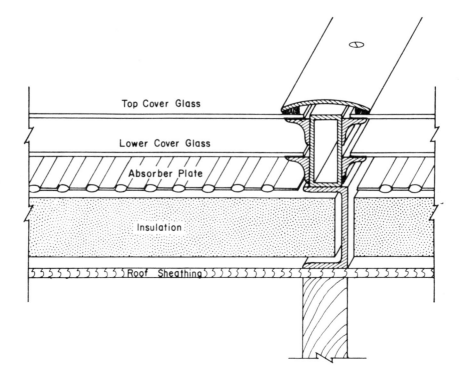

The use of aluminum for the structural components and as the absorber plate was predicated primarily on the basis of the metal's availability, durability, and cost. While copper has numerous advantages, its much higher cost appears to preclude widespread use. Steel is an alternative but is not currently available in the configurations desired. Low cost plastics were rejected because of their inability to withstand the temperatures of a no-flow situation (approaching 200°C).

A particular advantage of an aluminum absorber plate over other metals such as steel is its availability with preformed internal tubes. This type panel contains tubes and headers formed within the aluminum sheet. Thermal conductivity between the plate surface and the fluid circulated through the tubes is therefore maximized.

Selection of a tube pattern (see Figures 9.13 and 9.14) is based on two concerns: drainage and rate of flow (which affects heat transfer rate and pressure drop). Provision for drainage of the solar collector during nondaylight hours can alleviate freezing problems and eliminate heat loss from the water in the collector at night.

FIGURE 9.13: COLLECTOR PANEL TUBE PATTERN

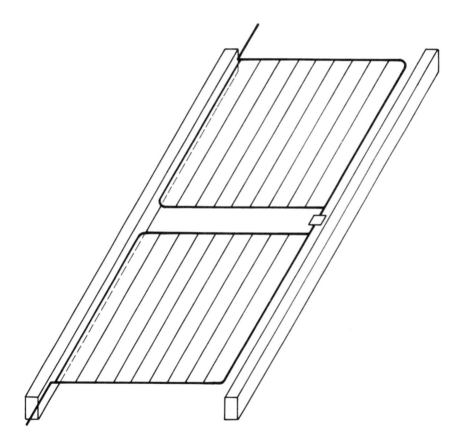

FIGURE 9.14: ALUMINUM TUBE IN SHEET DESIGN

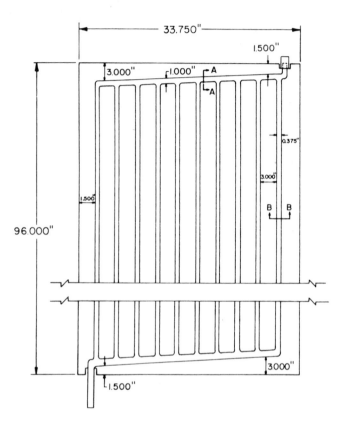

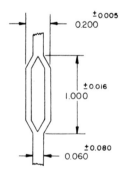

Section A-A

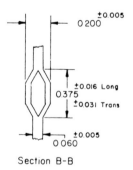

Section B-B

The rate of flow was designed to produce a 8°C (15°F) maximum temperature rise in the heat transfer fluid as it passes through the collector. For conditions of 1.36 l/min (300 Btu/hr/sq ft) and a solar collector efficiency of 40%, the flow is 227 l/sec (1 gpm) per panel [two 1 meter (3 feet) by 2.4 meters (8 feet) absorber plates in series]. Thus for all 16 panels the total flow to and from the collector is 3,632 l/sec (16 gpm). A 60:40 (by weight) mixture of ethylene glycol (with corrosion inhibitors) and water was used.

The aluminum absorber plate is coated with a flat black, high absorptivity paint. The 3M Nextel black coating has a total reflectance (visible to 35 microns) of less than 2%. Application of the paint required a thorough cleaning of the absorber plate, the use of a primer, painting, and subsequent baking at 200°C (400°F) to prevent any possible off-gasing of the paint when the collector became operational.

Determination of the solar collector design factors (collector slope from the horizontal, total area of the collector, and number of transparent covers) was carried out by the University of Wisconsin with a computer program by which the effects of these collector variables on the percentage load carried by the auxiliary fuel source were determined. Since the advantage of a solar system is the saving of fuel and not the elimination of the normal costs of a conventional heating and cooling system, auxiliary fuel use provides an excellent gauge of the solar collector characteristics.

Numerous possibilities were considered for the transparent panes, including glass, fiber glass reinforced polyester (FRP), and other plastics. Some plastics (in the price range of glass) cannot withstand typical weather conditions for more than a few months. Additionally, most plastics are partially transparent to the infrared reradiation from the absorber plate, making them less effective in controlling heat loss.

The optical diffusing characteristic of some plastics can be a disadvantage in double glazed collectors. Tests with an Eppley pyranometer showed a particular glass sample and an FRP plastic each had a solar transmission loss of about 13%. While two sheets of glass accounted for a loss of about 21%, a system comprising a layer of plastic above a glass layer suffered a 27% transmission loss. In view of the uncertainties in durability and other properties of plastics, it was decided that glass would be used on all surfaces. In this way, known liabilities are accepted in lieu of uncertain advantages.

Glass has an advantage, as well, in the possibility of being treated to reduce reflection losses. Several panels of the Colorado State University solar collector contain glass treated by Honeywell, Inc. to reduce reflection losses (normally 8% at normal incidence for glass with a refractive index of 1.5). This was accomplished by a process originally developed by RCA requiring dipping the glass in hydrofluosilicic acid. The surface should be durable inasmuch as glass treated 15 years ago by this process and exposed to the weather since then shows 96% solar transmission.

Of the 16 panels in the solar collector 4 contain glass which has been treated to reduce the reflection loss from 8% to approximately 5%. In addition 5 panels contain glass that was partially treated (reducing reflection loss from 8% to about 6%). The remaining panels, with one exception, contain untreated glass. The exception is a collector panel with an infrared reflective coating on the interior of the glass (developed by Libbey-Owens-Ford). The object here is to reflect the infrared reradiation from the absorber panel instead of absorbing this reradiation. While the RCA treatment process attempts to increase the amount of solar energy incident upon the absorber plate, the reflective coated glass is designed to lessen the heat loss by reradiation.

Each sixteen-foot panel contains six sheets of glass, each sheet measuring 0.86 meter (34 inches) wide by 1.37 meters (54 inches) long. These sheets were placed in a configuration of double glazing and three sheets along the sixteen-foot panel length.

Experience has tended to indicate that imperfections in the edge of the glass have a strong tendency to cause cracking and subsequent breaking in the interior glass sheets due to thermal stressing. Consequently, the edges of each sheet were sanded along the entire perimeter to remove these slight imperfections. In addition, glass that was chipped from shipping or handling was discarded. However, one panel was loaded with glass that contained some chipped edges as a comparison check.

Each sheet of glass was supported on all four sides by structural aluminum, and held in place by a butyl tape. Utilizing a primer, the butyl tape adhered quite well to the aluminum. The adhesion to the glass, however, was insufficient to prevent some creep of the interior glass at temperatures over 120°C (250°F). This creep amounted to about 20 cm (8 inches) over a period of three days when the collector panel was tilted at an angle of 45° from the horizontal. Small braces were added to the cross bracing to prevent subsequent creeping.

While some of the structural aluminum was cut and preassembled in the shop, the main components of the collector were installed on the subroof. Aluminum channels were first laid down and bolted to the subroof. Small Teflon strips were inserted at each bolt connection to allow for differences in thermal expansion between the aluminum collector and wooden subroof. Each collector panel was then assembled one at a time.

Insulation was placed between the aluminum channels first, followed by the absorber panels. Plumbing connections between the panels and the manifolds were made. The butyl tape was attached to the aluminum structural components, and each sheet of glass was then set in place. An aluminum cap strip was then attached to assist in holding the glass in place.

This method of on-site construction proved to be quite troublesome and labor-intensive. In addition to bad weather considerations, glass handling on a steep roof, working across a three-foot span on a partially completed panel, and the movement of tools and personnel to and from the subroof added considerably to the labor costs of the project.

One advantage of the particular installation method was the venting capability. As each of the collector panels was completed in place, the ends of the collector panel were left open. This allowed a natural convection through the panel and prevented excessive temperatures in the collectors during the installation. The ends were then closed following start up of the collector pumping system.

Due to the possibility of pump or power failure, and possible subsequent vaporization of the collector fluid, an expansion tank was added to the collector loop as a pressure relief component. A schematic of the expansion tank and collector loop is shown in Figure 9.4. The tank will serve to receive any collector vapor or liquid due to a boiling condition, and prevent excessive loss of collector fluid.

COOLING SYSTEM

LiBr Absorption Cooling Unit

The criteria for the selection of a subcooling unit involved immediate commercial availability, compatibility with solar heat supply, capability to meet cooling requirements, proven performance, acceptable cost, and adaptability to various climates on a national scale. For the Colorado State University solar laboratory the system selected was an Arkla-Servel 3-ton lithium bromide absorption cooling unit, modified to utilize hot water (instead of natural gas) as the heat supply to the generator. The Arkla gas-fired unit was initially rated as a 3½ ton unit. As a solar unit, it is rated at 3 tons.

A schematic drawing of the air conditioner is shown in Figure 9.15. In a typical absorption system, a solution of refrigerant and absorbent, which have a strong chemical affinity for one another, is heated in the higher pressure portion of the system (generator). This drives some of the refrigerant out of the solution. The hot refrigerant vapor is then cooled until it condenses and can be passed through an expansion valve into the low pressure portion of the system.

The reduction in pressure through this valve facilitates the vaporization of the refrigerant in the evaporator, where heat is removed from the environment. The vaporized refrigerant is next recombined with the absorbent mixture from which it was initially obtained to form a solution which is rich in refrigerant. The mixture then moves back into the high pressure side of the system where it is again heated and the cycle continues.

The generator in a gas-fired Arkla machine was replaced with one supplied by hot water entering at the bottom. With 38 l/min (10 gpm) of cooling water at 24°C (75°F), a hot water supply of 42 l/min (11 gpm) at 87°C (188°F) will provide the full 3-ton capacity.

FIGURE 9.15: SCHEMATIC FLOW DIAGRAM OF AIR CONDITIONING UNIT

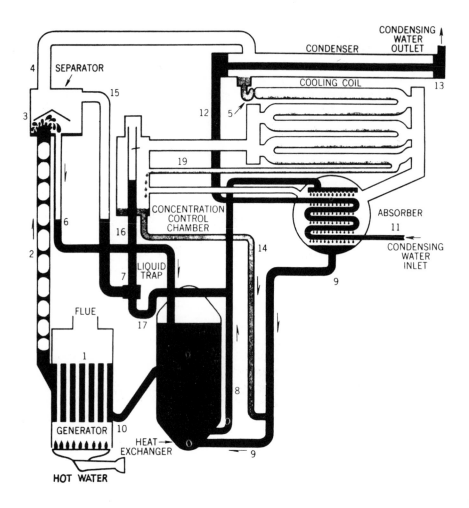

At lower hot water temperatures, cooling capacity is lower, reaching 77% at about 81°C (178°F). Design temperature limits for hot water supply are 80° to 95°C (175° to 202°F).

The temperature inside the evaporator is in the range of 3° to 10°C (35° to 50°F) (direct expansion). While it is possible to lower this temperature and reduce evaporator and condenser surfaces, this would increase the chances of freezing and is therefore not recommended. While lower cooling water temperature permits a lower evaporator temperature, cooling capacity is not appreciably affected. Rather, early vaporization occurs, and in the case of very low cooling water temperatures, freezing in the evaporator may result. In order to take advantage of the lower cooling water temperature, the concentration of the LiBr solution must be reduced.

Because of cooling water temperatures typically about 5°C (10°F) lower than the 24°C (75°F) design level, heat supply temperatures can be about 3°C (6°F) lower than usual. To account for both these differences, the concentration of the LiBr solution has been reduced to 51 to 54%, from the customary 54 to 57%. Temperature-concentration diagrams show that crystallization cannot occur in the absorber or in the generator provided that generator input temperature does not fall below 80°C (175°F). Below this temperature vapor formation occurs at a rate too low for effective lifting of the LiBr solution in the bubble pump, so after a time, the concentration could increase to the crystallization point. The generator will therefore always be provided a heat supply at a temperature no lower than 80°C (175°F).

The unit has been modified also by the installation of a diverter valve for the cooling water to the absorber. The valve will divert approximately 50% of the cooling water when its temperature falls below 18.3°C (65°F). The valve remains open until the cooling water temperature rises to 20.5°C (69°F), giving an effective dead band of 2.2°C (4°F). It is noteworthy that the effect of the bypass valve is to increase the capacity of the cooling unit described by approximately 3,000 Btu/hr at the design generator temperatures. However, this advantage is lost when the input temperature to the generator is near its lower limits.

These modifications permit advantage to be taken of prevailing wet bulb temperatures, and consequently, allow an increase in capacity over most of the range of generator temperatures. The location of the Colorado State University solar house at 1,585 meters (5,200 feet) elevation, puts an upper limit of 95°C (202°F) on hot water temperature (local boiling point) if a nonpressurized system is used. The unit will, however, operate satisfactorily if boiling water is entering the generator as the vapor will provide heat by condensation.

Water Cooling Tower

In order to match the capacity of the Arkla 3-ton air conditioner, the cooling tower must be capable of cooling 10 gallons per minute of 93°F to 75°F at an outdoor wet bulb design condition of 65°F. The cooling tower must therefore have a capacity of about 90,000 Btu/hr with a 10°F approach and an 18°F range.

Of the cooling towers manufactured by the Marley Company, the closest in capacity was Model 4615 with 8 nominal tons (96,000 Btu/hr) of capacity. The air fan is driven at 668 rpm by a V-belt drive connected to a 115 volt, one-third horsepower, single phase, one speed, electric motor with a full load rpm of 1,735 at 60 Hz. The shipping dimensions of the tower are approximately 31 inches wide, 61 inches long, and 55 inches high. The tower is designed for water flow rates of 16 to 27 gallons per minute. Because the required flow rate for the Arkla unit is 10 gallons per minute, different nozzles were required. The tower weighs 410 pounds and includes an air inlet screen and hot water basin cover. Steel extensions were bolted to the four short (three and three-quarter inch) legs in order to make use of the two-inch diameter outlet on the very bottom of the tower. The galvanized steel tower has a noncombustible asbestos fill.

A Penn Aquastat water cooling tower electric fan control permits operation of the cooling tower fan as a function of cooling tower discharge water temperature. There are two conditions, both of which must be met, for the cooling tower fan to operate. First the Arkla unit must be on, and second, the water temperature from the cooling tower must be at least 75°F.

To protect the cooling tower from freezing, the line from the bottom of the tower discharges into a small tank in the basement. This discharge line terminates below the minimum liquid level in the indoor water tank to avoid the excessive noise associated with a discharge above the water surface. The manufacturer recommends that the capacity in gallons of this inside open-type tank be four times the flow rate through the tower in gallons per minute.

The indoor tank is a common 55-gallon drum with the top cut out. The cooling tower pump suction is connected to the bottom of this drum and pumps directly to the Arkla unit. After passing through the air conditioner, the warmed water goes to the top of the cooling tower. A small valve on this line is adjusted to give the proper blowdown when the unit is in operation.

At the top of the indoor cooling water tank is a three-quarter inch float valve which provides the make-up water needed. Just below this is an overflow line leading to a floor drain in order to prevent the water in the indoor storage tank from entering the make-up water supply line, should the make-up water system pressure fail for some reason. The temperature sensor of the Penn Aquastat cooling tower fan motor control is submerged in the tank. The above described installation of the cooling tower eliminates all freezing problems. It is thus possible to have subfreezing temperatures at night and also to have refrigeration cooling during the day.

ECONOMIC CONSIDERATIONS

In the process of constructing the solar system and its installation in the house, detailed cost accounting was performed. The results are associated with the following applicable factors. The unit is a custom design of comparatively small size. In evaluating the necessary man hours for repetitive tasks, the time requirement for the last units was used so that learning time would not be included. University shop personnel, normally accustomed to new and unique work assignments, did most of the assembly and installation. Tooling costs were minor and were not included. Extra costs are associated with the research aspects of the facilities, including redundancy in operating methods.

Since the cost of solar heating and cooling is the most important consideration in its adoption and practical use, there must be a clear understanding of the criteria.

First, it is necessary to separate the cost of the solar components from the cost of the house itself. The cost of the house should be considered to include a conventional heating and cooling system, capable of meeting maximum demands. The solar energy system cost is then the additional costs of the solar components, such as the solar collector, thermal storage, associated piping, valves, etc. This does not include furnace, distribution ducting, an air conditioner, water heater, and conventional controls because they would be used in a fuel operated system, and they are also the conventional parts of the solar operated system. The air conditioner might be of a cheaper type in a conventional system, however.

While the solar system could be built so as to meet the complete heating and cooling requirements of the building, this would be very uneconomical. Rather, the solar system should be used as a supplementary energy source (but perhaps the major part) and serve only to save fuel used in the conventional heating and cooling unit. To appraise the economics of a solar energy heating and cooling system with a conventional system, the fuel cost savings are compared with the solar costs. For simplicity, it is assumed that the solar costs are covered by a loan which is amoritized over a period of time at a specified interest rate. In this way one can easily compare the monthly costs of the solar system with conventional fuel saved.

The solar system in the Colorado State University solar house was designed to handle about three-fourths of the heating and cooling load of a typical house of 3,000 square feet. The total cost of the solar components (including shipping, installation, and associated costs) was approximately $6,500. Of this the costs of the solar collector were about $5,000. At 1974 interest rates this could constitute a monthly payment of about $60 in order to save an expected three-fourths of the heating and cooling fuel bill.

DATA ACQUISITION AND HANDLING

Major emphasis has been placed upon collecting complete and accurate performance data. The performance of the solar heating and cooling system and its several principal components, and the influence of environmental and operational conditions on such performance are the essential elements in evaluation of the design. There is also the need for correlating performance with solar input and load factors so that design improvements can be introduced. Interaction between components in the system and the continuous variation in operating conditions require the measurement of many variables on a continuous basis.

The Colorado State University solar house has been equipped with a data acquisition system to gather 74 channels of data on climatic conditions, building load, and equipment performance. The system is expandable to 100 channels and records up to 2 channels per second. This data acquisition capacity is a powerful asset to a solar installation with experimental emphasis. By its use, numerous individual effects, such as those obtained by use of antireflective cover glass treatment, may be analyzed apart from the total operation.

The computer program converts the various analyses to desired values and compiles them from the data. At the end of each hour, values are printed, and daily totals are printed out at solar midnight. The information which describes total system performance will be processed on a one or two day interval by transfer from the magnetic tape files into the Colorado State University CDC 6400 for computation. The tape files will be stored for future reference and for more detailed analyses.

Other Ongoing and Proposed Research

SCHOOL BUILDINGS

Timonium Elementary School—Baltimore County, Md.

On March 7, 1974 the Timonium Elementary School in Baltimore County, Maryland became the first school in the nation to utilize solar energy for classroom heating. AAI Corporation, a subsidiary of United Industrial Corporation, announced start-up of the solar heating system installed to heat one wing of this three-wing building. Five thousand square feet of collector panels will heat water to replace the existing oil-fired steam heat system in this wing of the building. Figure 10.1 illustrates the installation. The solar heating experiment is a RANN project jointly sponsored by the National Science Foundation and AAI. Tests made during the first days of March indicate that the efficiency of the system is at least as high as originally planned.

The system design is based on a solar collector panel which serves the dual purposes of collecting solar heat and of roofing the building. In the school project the panels are installed over the existing roof, but in new building construction, the solar panels can be used as the complete roof structure itself. Thus installed, they will withstand all normal wind, snow and building loads. Structural testing of the panels at temperatures from $0°$ to $200°F$ has been successfully accomplished.

AAI Corporation is proposing an addition to the Timonium School project that will provide complete summer air conditioning by solar energy. No additional collector panels will be needed and the dual system will make the complete project economically feasible for other schools that need both winter heating and summer cooling.

FIGURE 10.1: TIMONIUM ELEMENTARY SCHOOL

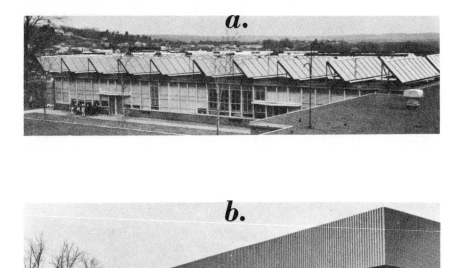

Attractive Appearance After Installation

Source: AAI Corporation, Baltimore, Md.

Grover Cleveland Junior High School—Dorchester, Mass.

Solar energy is more meaningful to students at Grover Cleveland Junior High School in Dorchester, Mass. since a General Electric Company experimental solar heating system, funded by the National Science Foundation, is providing a portion of the school's heat.

GE has modified the school's conventional electrical heating system to accept supplementary solar heat. About 4,500 square feet of GE experimental solar heat collector panels have been installed on top of the school roof in three rectangular rows. The system contains approximately 150 panels, each measuring 4 by 8 feet. The solar panels consist of a black, heat absorbing surface beneath two rigid sheets of clear Lexan plastic. A tubing network inside the black surface is filled with a water/antifreeze solution. When energy from the sun passes through the plastic, it is absorbed by the dark surface and converted into heat, which in turn is transferred to the liquid. The hot liquid is then pumped through the tubing network to a pair of special solar heat exchangers that work in conjunction with two of the school's ten conventional heating units. These solar heat exchangers then heat the air which warms a portion of the building as needed to maintain a comfortable temperature.

Whenever the system generates more heat than is required, surplus hot water will be stored in a 2,000-gallon heat storage tank. From here, the hot water can be pumped to the solar heat exchangers providing heat to the classrooms whenever clouds obscure the sun and the solar panels are not absorbing heat.

The experimental GE system is designed to provide up to 20% of the heat required to warm the three-story school building. This percentage will vary according to such factors as outside temperature, cloudiness, relative humidity and time of day. Major GE subcontractors on the Grover Cleveland project include the Ballinger Company, Architect Engineers, Philadelphia, Pa. and the Vappi Company, Cambridge, Mass.

Figure 10.2 depicts a cross section of a General Electric solar energy collector panel, as used on the experimental solar energy heating project. About 150 panels, covering approximately 4,500 square feet have been installed on the school roof. These panels are collecting solar energy and converting it into heat for warming a portion of the three-story building.

FIGURE 10.2: SOLAR COLLECTOR CROSS SECTION

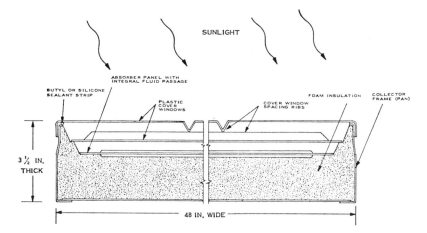

Source: General Electric Company, Philadelphia, Pa.

As sunlight passes through a panel's clear plastic cover windows, the solar energy is retained by the windows, absorbed by a dark surface, and converted into heat.

This heat is transferred to a liquid, which then is pumped through an integral fluid passage tubing network in the panels to a pair of special solar energy heat exchangers. The heat exchangers then provide the heat needed to maintain a comfortable temperature in a portion of the school building. Figure 10.3 depicts the experimental solar heating system employed by Grover Cleveland Junior High School to provide up to 20% of the school's heat.

FIGURE 10.3: SCHEMATIC OF OPERATIONAL MODES

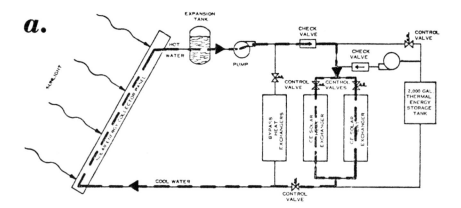

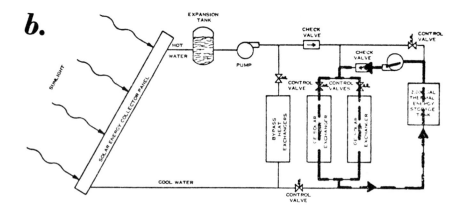

(continued)

FIGURE 10.3: (continued)

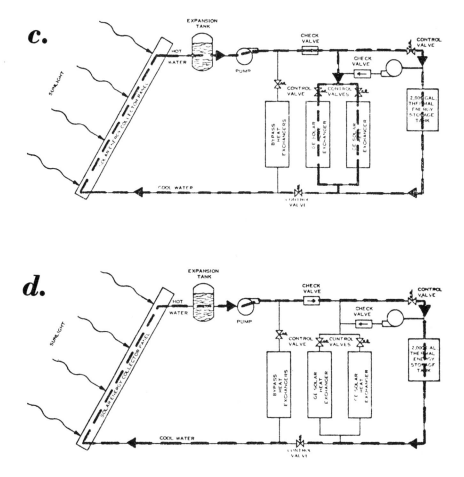

Source: General Electric Company, Philadelphia, Pa.

Figure 10.3a depicts the school rooms using energy provided by solar energy collector panels. This mode is used whenever the amount of heat collected by the panels equals the amount needed in the classrooms. Figure 10.3b shows the heating of school rooms using stored energy. This mode could be used for operation during early morning, evening or night periods or during cloudy weather.

Figure 10.3c shows the heating of the school and also heating of the thermal energy storage tank. These two modes could be employed when more heat is being collected by solar panels than is required for heating classrooms. Figure

10.3d shows the heating of the thermal energy storage tank. This mode could be used when no heat is required in the classrooms.

North View Junior High School—Osseo, Minn.

This is a Honeywell installation in suburban Minneapolis which according to *Commerce Today,* vol. IV, No. 12, March 18, 1974, uses solar thermal power for heating. The system is expected to save 12,500 gallons of fuel oil annually.

Fauquier County Public High School—Warrenton, Va.

According to *Chemical and Engineering News,* January 21, 1974, the Fauquier County Public High School will be equipped with a solar heating installation, under an NSF contract, by Inter Technology.

George A. Towns Elementary School—Atlanta, Ga.

The Towns Elementary School in Atlanta is scheduled to undergo a retrofitting of its heating system to convert it to solar heating and an installation of a solar cooling system. Solar energy is expected to supply 60% of the school's energy needs. Ten thousand square feet of flat plate collectors will be roof-mounted and hot water will be the circulating medium. As reported by R. Duncan in *ASHRAE Journal,* September 1974, the storage system will be four 6,000-gallon tanks. No long-term energy storage is envisioned. Systems evaluations on this project will be carried out by the Georgia Institute of Technology in cooperation with Westinghouse Electric Corp.

Denver Community College—Westminster, Colo.

Denver Community College at Westminster, Colorado is one of the largest buildings planned in the United States so far according to F.H. Bridgers, *ASHRAE Journal,* September 1974. The 278,800 square foot solar-heated building, to house 3,500 students, is to be completed in 1976. The building will have 50,000 square feet of collectors, a 400,000-gallon reservoir for water which will circulate through the system and heat pumps for use in summertime air conditioning and auxiliary winter heating. The solar designing for the building was done by Bridgers and Paxton Consulting Engineers, Albuquerque, New Mexico.

RESIDENTIAL BUILDINGS

Solar Residence—Shanghai, W.Va.

The project described by R. Rittleman in PB-223 536, *Proceedings of the Solar Heating and Cooling for Buildings Workshop, Held in Washington, D.C. on March 21-23, 1973,* R. Allen, editor, prepared for the National Science Foundation, issued July 1973, is a demonstration residence to be located in

Shanghai, West Virginia, approximately six miles southwest of Martinsburg.

The residence (Figure 10.4) contains approximately 1,400 square feet of finished living space and approximately 350 square feet of mechanical equipment space. The project will contain the following subsystems: a solar thermal collection system (Figure 10.8), a photovoltaic conversion system, a wind driven electrical generating system, and an aerobic composter for organic wastes.

The first two subsystems are intended to provide a platform for the ongoing evaluation of various manufacturing techniques, component designs, and comparative data. They will be designed and installed in such a manner as to facilitate flexibility and ease of installation and replacement of the various components. The latter two subsystems are included primarily for demonstration purposes.

The basic purpose of the project was to show the integration of innovative mechanical systems with contemporary and functional architectural design in contrast to the "laboratory" character of past projects of this nature. The architectural solution, however, is unique in that it responds to the severe energy conservation requirements of a solar system.

The usual high heat losses associated with extensive glazing have been minimized by using a technique of triple glazing, using a standard insulating glass to the exterior, a three-foot air space and a single operable glass on the interior, the space between being used for house plants. The exterior door most often used opens to the east and into a small greenhouse area which serves as an air-lock to minimize infiltration losses. Voids have been provided at the roof edges and at the roof peak to facilitate access to plumbing and air duct connections for ease of installation, testing and replacement with alternate components.

The main solar array shown in the south elevation view of Figure 10.4 is 588 square feet in area and is positioned 180° in azimuth and 45° to the horizon (latitude +6 degrees). The collector units are designed to be integrated with the structural system on a 32" center to center module. The individual collector modules will be 96" long thus affording a packageable unit that is light in weight and can be installed by two men.

The collector unit construction consists basically of a flanged 0.040" metal pan with 2" of fiber glass insulation on the bottom. This will be covered by the heat transfer coil array. The array will be 0.25" diameter aluminum tubing at 6" spacing on a 0.040" thick aluminum plate. The entire plate will be sprayed with a commercially available flat black paint. The unit will be covered by a ⅛" thick window glass (DS).

A 50% solution of ethylene glycol will be the heat transfer fluid and will be mechanically circulated in the loop shown in Figure 10.5 at a rate of 8 pounds per square foot per hour. This flow rate will be variable to allow for evaluation of various assemblies at optimum flow conditions.

FIGURE 10.4: SOUTH ELEVATION

Source: PB-223 536, July 1973

FIGURE 10.5: SOLAR SYSTEM SCHEMATIC

Key Shown on
Following Page

Source: PB 223 536, July 1973

Key to Figure 10.5

1) Return air (70°)
2) Supply air, 85° minimum, 125° maximum
3) Fan-coil unit, 660 cfm, 1,600 gpm
4) Fan-coil circ pump, 4 gpm
5) Solar collector circ pump, 12 gpm
6) Season changeover 3-way valve, manual
7) Auxiliary heater, 84,000 Btu output
8) Auxiliary heater circ pump
9) Season changeover 3-way valve, manual
10) Suction selection valve, 90° minimum
11) 24,000-gallon storage tank, 90° to 130°
12) 400-gallon storage tank, 135° to 150°
13) Domestic hot water immersion coil
14) Domestic hot water tempering valve
15) Cold water inlet
16) Expansion tank and collector receiver
17) Solar collector modules
18) Balancing valves

The cover glass will be placed on a glazing tape and the entire assembly will be held within the structural system with neoprene glazing gaskets. This will permit a weathertight seal and accommodates rapid, easy removal and replacement of various assemblies for testing.

The circulating pump will be controlled by a comparative thermostat which will energize the pump when the collector temperature exceeds the large storage tank temperature by 5° (for winter heating). For summer operation, the comparison will be made with the small storage tank.

The thermal storage system is a two temperature liquid, sensible heat tank located beneath the residence. A two temperature tank was selected to allow greater flexibility in adaptation to other subsystems, and allow the optimum use of available insolation.

Storage systems in most of the past attempts have tended to be expensive to be efficient. Attempts to use crushed rock for sensible heat storage have been reasonably successful at low cost but consume considerable volume and offer much resistance to air flow, increasing the fan horsepower needed to overcome this resistance. In addition, the design calculations become more empirical and the outcome less predictable. For these reasons, water was selected for the storage medium.

One of the objectives of this program is to investigate the cost effectiveness of providing a relatively large volume water storage system. Water storage in past examples has typically been large insulated steel tanks buried underground or surrounded by crushed stone. By constructing the tank using an extension of the already existing foundation walls of the residence, lining with common urethane insulation board and using a tank lining, low cost storage assembly should be obtained.

The total water storage capacity is 2,400 gallons (19,200 pounds) in the large tank and 400 gallons (3,200 pounds) in the small tank. The useful temperature of the large tank will be 90° to 130°F for winter heating and will afford a useful heat storage capacity of 768,000 Btu. This will provide heating for 41.3 hours at the January mean temperature of 33.9°F and for 21 hours at 0°F. The 2,400 gallon tank will be the main sump for the solar collector during the winter heating season. This tank will supply the fan coil unit if the tank temperature is 90°F or more.

The 400-gallon tank is heated by the auxiliary heater during the winter heating season and is used to provide domestic hot water and standby for house heating when the 2,400-gallon tank falls below 90°F. The useful temperature range of the 400-gallon tank is from 130° to 150°F and is maintained within this range by a hi-lo thermostat on the auxiliary heater. At maximum temperature, this tank will store 192,000 Btu of heat, enough for another 10.3 hours of heating at the January mean temperature, disregarding the loss for heating domestic hot water. The total system heat storage, therefore, provides for 51.6 hours of house heating at the lowest average winter temperature.

It is not anticipated that the temperature of the 2,400-gallon tank will exceed 120°F during the winter. It is possible, however, that during the changeover periods in early May and late September that the 2,400-gallon tank may reach capacity. At that time, the collector will use the 400-gallon tank as a sump and bring its temperature to 150°F or above. This avoids the use of any auxiliary heating as long as there is solar insolation available.

An oil fired water heater having an output of 84,000 Btuh was selected for auxiliary heating. The use of oil allows the project to store its own energy. If the gas or electric utilities were to be connected for auxiliary purposes, the use would occur when the utilities are least able to furnish energy and would serve only to aggravate the maximum demand conditions which the utilities are called upon to serve. The capacity of the heater is the smallest commercial unit available. This selection is possible because the heater is buffered from instantaneous demands by the 400-gallon storage tank. The auxiliary heater will be controlled by a 135° to 150°F hi-lo thermostat and will be called upon for heat only when there is insufficient solar insolation to maintain both tanks at maximum temperatures.

The heating of the house will be accomplished with a fan-coil unit having a variable speed blower motor. The motor speed varies with the coil inlet water temperature, ranging from 660 cfm at 130°F inlet temperature to 1,600 cfm at 90°F inlet temperature.

Other than the control of the variable speed blower motor, the house thermostat will operate the circulating pump which will be electrically interlocked with the blower motor for normal operation. The suction of the circulating pump will normally be the 2,400-gallon tank. This suction will automatically switch to the 400-gallon tank if the large tank temperature drops below 90°F.

Domestic hot water will be supplied through a tankless coil immersed in the 400-gallon tank. Auxiliary heat is anticipated in 3 winter months. The need for heat for domestic hot water is in addition to those requirements, and auxiliary heat will be used for those months for all domestic hot water. During the summer months, the collector will be used for maintaining the temperature of the 400-gallon tank for occasional house heating and all domestic hot water heating.

Provisions have been made in the design of the system to accommodate an absorption air conditioner in the future. The heat source for the absorption unit will be the 400-gallon tank. During the summer months, it will be possible to boost the temperature of this tank to approximately 180°F using the solar collector. If the need should arise, the tank could be boosted by the auxiliary heater as well as by changing the control limits on the hi-lo thermostat. When the 400-gallon tank is being used to power the absorption unit, it will be necessary to install a mixing valve on the outlet of the domestic hot water coil.

When the absorption unit is installed, the 2,400-gallon tank will be used as an evaporator and coolness storage. Using the absorption unit to pump down the 2,400-gallon tank temperature will allow the dampening of extreme daily cycling and minimize the required absorption unit capacity. The same fan-coil unit will be used for cooling with the controls reset for summer operation. Again, the variable speed blower motor will be controlled by coil inlet temperature.

Odeillo, France Dwellings

According to J.D. Walton, Jr. of the Georgia Institute of Technology (PB-223 536, *Proceedings of the Solar Heating and Cooling for Buildings Workshop, Held in Washington, D.C. on March 21-23, 1973,* R. Allen, editor, prepared for the National Science Foundation, July 1973), in addition to the three-unit dwelling described previously, construction has begun on 31 additional units.

All 34 units will be located near the 1000 kw solar furnace. These houses will utilize the current state of technology related to the use of solar energy for space heating. Some units will use water for heat storage. Various building materials and types of insulation will be evaluated in order to optimize various structures with respect to efficiency, economics, size of dwelling, collector area, etc. The Georgia Institute of Technology Engineering Experiment Station plans to arrange with the researchers of the Solar Energy Laboratory to collect and evaluate construction costs, efficiencies, economics and over-all capabilities of the dwellings being constructed at Odeillo.

Multifamily Dwelling—Hamden, Conn.

The first multifamily dwelling utilizing solar heating will be a 40-unit development for the elderly at Hamden, Connecticut according to the *New York Times*
December 31, 1974. Construction is scheduled to begin in July 1975.

All 40 units will have conventional heating but 20 of the units will have solar
heating in addition and will only use the conventional heating as a back-up
system. McHugh and Associates of Farmington, Connecticut are the architects;
Minges Associates, also of Farmington, are the consulting engineers and Kamen
Aerospace Corporation of Bloomfield, Connecticut will be the builder.

Arroyo Barranca Project

According to H. Barkmann of Sun Mountain Design, Santa Fe, New Mexico, in
PB-223 536, *Proceedings of the Solar Heating and Cooling for Buildings Workshop,
Held in Washington, D.C., March 21-23, 1973,* prepared for the National Science
Foundation, issued July 1973, an attempt is under way to design a 250-unit
housing project in which the units will be located in clusters on some 68 acres.

The purposes of the Arroyo Barranca project are twofold:

1) to develop a land-use plan to preserve the environment and
2) to use existing technology such as solar treating to complement
   the design of minimum environmental impact.

Much of the work has been based on studies of old Indian pueblo and
present Mediterranean villages. Typical house units are thick walled buildings
with low heat loss.

Figure 10.6 illustrates a view from the south. It shows the solar collectors
in an architectural rendering which suggests that such a system can be
attractive and need not be offensive. The houses are arranged in eleven-unit
clusters.

Figure 10.7 is a phantom view of a hot water system, depicting the solar
equipment, auxiliary equipment and heating panels in relation to the building
shell.

Figure 10.8 depicts a hot air system showing the ductwork for circulating
air through the solar collector.

A cost comparison of hot water and hot air systems was made, based on a
$4.00-per-square-foot collector. The hot air system cost was found to be
60% of the cost of the hot water system. The thrust of the project is that,
for environmental reasons, existing solar technology will be used to build now,
and allow future refinements and developments in solar technology to improve
future projects.

FIGURE 10.6: VIEW OF HOUSE UNITS FROM THE SOUTH

Source: PB-223 536, July 1973

FIGURE 10.7: HOT WATER HEATING SYSTEM

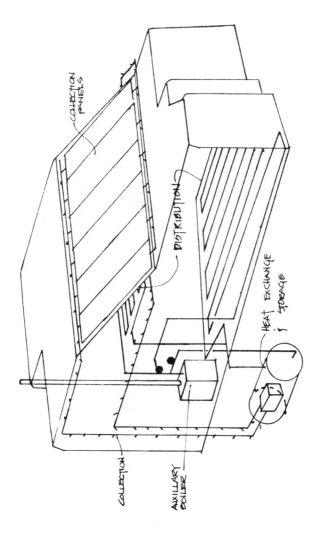

COLLECTION PANELS

DISTRIBUTION

HEAT EXCHANGE & STORAGE

COLLECTION

AUXILIARY BOILER

Source: PB-223 536, July 1973

FIGURE 10.8: HOT AIR HEATING SYSTEM

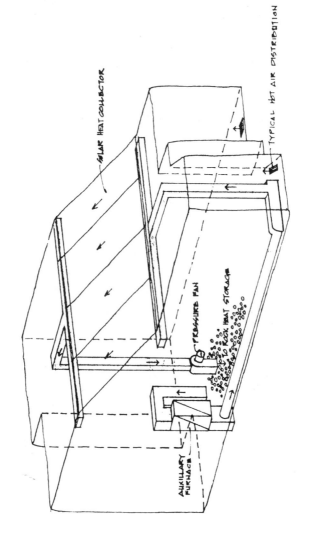

SOLAR HEAT COLLECTOR

TYPICAL HOT AIR DISTRIBUTION

PRESSURE FAN

TO ROCK HEAT STORAGE

AUXILIARY FURNACE

Source: PB-223 536, July 1973

Project Ouroburos—Minneapolis, Minn.

A two-story 1,500 square foot dwelling at 45°N latitude built by University of Minnesota School of Architecture and Landscape Architecture students under D. Holloway in 1973 has been reported to have a 12-panel, 576 square foot solar collector made of corrugated galvanized sheet metal with two sheets of glass cover (*Environmental Action Bulletin,* September 22, 1973). Thermal storage is as sensible heat employing a 1,250-gallon insulated water tank and a 1½ horsepower circulation pump for transfer. The collector is in the form of a south-facing wall. Auxiliary heating is provided by an electric immersion heater and/or a wood-fired stove. Future use of heat pumps is contemplated.

A second project by this group was the conversion of a two-story 2,000 square foot house from conventional to solar heating. The solar system consists of a 600 square foot rooftop collector at a 45° tilt and a 500 square foot south-wall collector. Heat storage is in a 2,000-gallon basement water tank. The collector to storage loop is water; the storage to space loop is air.

Dome Home—Albuquerque, N.M.

This home, built in 1972 by R. Reines, in the shape of a 31½ foot diameter dome, contains 855 square feet of usable space on two levels. The conventional flat plate collector is separate from the house on a nearby slope. (A second, similar-sized dome built subsequently as a workshop has its collectors integral with the structure.) Thermal energy is derived from the sensible heat of water in a 3,000-gallon polyurethane foam-insulated tank which is also separate from the house.

The house, located at a latitude of 35°N, achieves 100% space heating and water heating from the sun. Heat transfer from collector to storage is via water-glycol and from storage to space via water. The system has a storage capacity sufficient for seven sunless days according to the *Los Angeles Times,* January 1, 1973.

Grassy Brook Village—Brookline, Vt.

A condominium under construction in Brookline, Vermont will employ a single 4,500 square foot solar collector for a ten-unit group of houses. The collector, at a tilt of 57°, will be arranged in saw-tooth fashion to reduce height and wind loads and will be located separate from the houses. This will leave the south-faces of the houses free for use as large windows to allow direct solar heating when desired. These windows can be covered with insulating shutters at night or during periods of extreme summer heat.

Grassy Brook Village has been developed by Richard Blazej, the architects are People/Space Co. of Boston and the engineers are Dubin-Mindell-Bloome. Completion of the project is expected in 1975.

OTHER BUILDINGS AND MODELS

State of New Mexico Department of Agriculture Building

The Department of Agriculture Building at New Mexico State Univeristy, Las Cruces, N.M. according to F. Bridgers, *ASHRAE Journal,* September 1974, will be about 25,000 square feet in area and will house laboratories using exhaust hoods. Solar heating facilities only will be installed initially as preliminary studies and calculations indicate that solar cooling would require three times the collector area and greatly increased costs.

NASA Langley Office Building—Hampton, Va.

As reported in N74-34541, *Solar Energy to Heat and Cool a New NASA Langley Office Building,* a NASA Technical Memorandum by W.L. Maag, Lewis Research Center, Cleveland, Ohio, issued September 1974, the NASA Lewis Research Center is investigating the use of solar energy to provide a portion of this nation's energy requirements.

In a joint effort with the NASA Langley Research Center, a solar heating and cooling system will be installed at a new office building to be constructed at Langley. The objective of this project is to establish a full-scale working test-bed facility to investigate solar energy for heating and cooling buildings. Results will provide technical and operational information required to establish solar collector performance and assess the use of solar energy for this application.

The building heating and cooling system will operate on $160°$ to $220°F$ hot water generated from solar energy or from a steam utility system. The solar system will initially be designed for 15,000 square feet of collector surface area, located directly adjacent to the building. Analysis shows that the system could perform at an over-all thermal efficiency between 30 and 50%. The energy collected at these conditions will provide between 80 and 100% of the heating and cooling requirements during the cool months and between one-half and two-thirds of the cooling requirements in the summer. Thermal energy storage will be provided to bridge the gap between cloudy days and clear days.

NASA plans to provide a solar heating and cooling system for a new office building currently scheduled for construction in 1974 and 1975. A solar hot water system will be installed as a part of the building's conventional heating-ventilating-air conditioning (HVAC) system, and is scheduled to be operational during 1975.

The objective of this project is to establish a full-scale, working test bed facility to investigate solar energy systems, particularly in areas where NASA has experience, such as thermal design, selective coating technology, and thermodynamic energy systems. The solar system will be capable of accepting up to 16 different types of solar collectors, thus serving as a full-scale test bed for both state-of-the-art and advanced solar collector technology.

The Langley System Engineering Building (SEB) was chosen for this project. It will be a single story structure enclosing 53,000 square feet of floor space for 350 engineering personnel. The building was originally designed by Langley with a conventional HVAC system using compression-cycle air conditioning. As part of NASA's energy conservation program, the Langley design includes an activated charcoal filtering system that permits the ventilation air requirements to be reduced 75% to 5.5 cubic feet per minute per person which reduces the air conditioning load about 10%.

To accept solar energy, the compression-cycle air conditioner will be replaced with an absorption-cycle air conditioner which can operate with hot water supplied by the Langley utility system or generated from solar heat. An area adjacent to the building (Figure 10.9) will be used to place the solar collectors, storage tanks, and piping and control system. The heating and ventilating and air conditioning (HVAC) system is illustrated schematically in Figure 10.10.

Ventilation air at $60°F$ is circulated through the building at a rate of 5.5 cubic feet per minute per person. This air receives the heat load produced by people, lights, transmission, infiltration, etc., and is either partially exhausted to the outside or recirculated through charcoal filters for reuse. When the outside air temperature is above $60°F$, the ventilation air is cooled in a heat exchanger that is supplied with chilled water at $45°F$ from the absorption chiller. Heat losses from the building surfaces during the winter months are balanced with heat supplied from baseboard heaters. The building is zoned for optimum utilization of heating.

The HVAC system can be operated with hot water produced either from solar collectors or from a heat exchanger operated with steam from the Langley utility system. The energy is utilized in two stages. In the temperature range between $200°$ and $240°F$, the hot water operates an absorption cycle (lithium bromide-water) chiller that will produce 50 to 150 tons respectively of cooling in the form of chilled water to cool the ventilation air.

The hot water leaves the absorption unit at a minimum temperature of $180°F$ which is sufficient to then provide heat to the building through the baseboard convectors before it is returned to the heat source at a minimum temperature of $140°F$. The system can operate separately to provide either heating or cooling, and simultaneously for both.

The 15,000 square foot collector area will provide most of the heating and cooling requirements during the late fall, winter and early spring months of the year, and between one-half and two-thirds of the cooling requirements during the summer. Storage will bridge some of the gaps between cloudy and clear days. Therefore, this size should be sufficient to demonstrate operation of the solar heating and cooling system for all modes of operation, i.e., heating only, cooling only, heating and cooling at the same time and switching from one mode to another. If additional area is desired in the future, space is available to expand the collector field to 40,000 square feet of collector surface area.

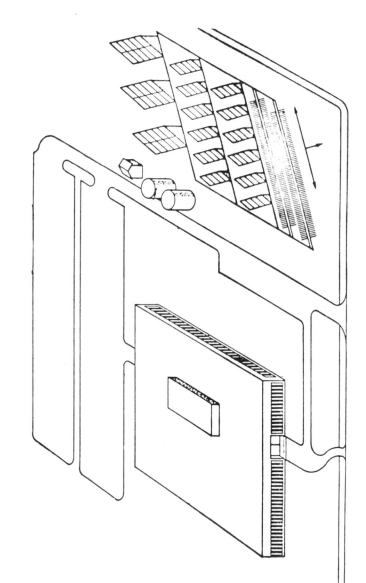

FIGURE 10.9: LANGLEY OFFICE BUILDING AND COLLECTOR SYSTEM

Source: N74-34541, September 1974

FIGURE 10.10: SCHEMATIC OF COLLECTOR SYSTEM

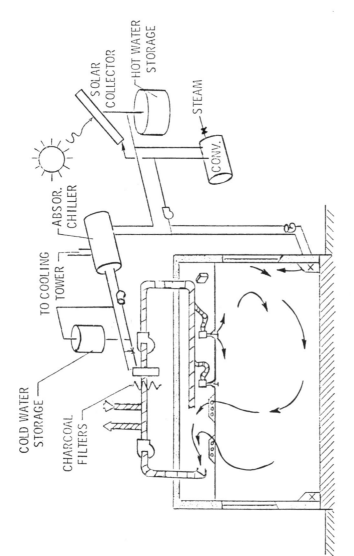

SOLAR COLLECTOR

HOT WATER STORAGE

STEAM

CONV.

ABSOR. CHILLER

TO COOLING TOWER

COLD WATER STORAGE

CHARCOAL FILTERS

Source: N74-34541, September 1974

Somerset County, N.J. Park Commission Building

New Jersey's Somerset County Park Commission according to the *New York Times,* December 17, 1974 is building an Environmental Education Center in Basking Ridge, a 9,000 square foot structure which will be solar heated using collectors, consisting of serpentine pipe located beneath layers of glass and Plexiglas located on a south-facing roof. Heat will be circulated via 6,000 gallons of water from underground storage tanks. The building was designed by Halsey and Ryder, architects, of Basking Ridge and Becht Engineering Co. of Denville, N.J. Completion is scheduled for 1975.

Massachusetts Audubon Society Solar Building—Lincoln, Mass.

According to J.C. Burke of Arthur D. Little, Inc. and the Massachusetts Audubon Society (PB-223 536, *Proceedings of the Solar Heating and Cooling for Buildings Workshop, Held in Washington, D.C., on March 21-23, 1973,* R. Allen, editor, prepared for the National Science Foundation, July 1973), the Massachusetts Audubon Society is considering a solar space-conditioned building for additional office space at its headquarters in Lincoln, Mass.

The construction of a solar building , by involving industry and government in the design and construction, will give Audubon an opportunity to contribute to the acceptance of solar energy for building space conditioning, and at the same time, satisfy a need for additional office space at their Lincoln, Mass., headquarters. MAS can add to the credibility of such a solar demonstration because of their background as a major conservation organization devoted to the public interest.

The Massachusetts Audubon Society hopes to accomplish a number of objectives in the design, construction, and operation of this solar building. These objectives include demonstrating the feasibility of solar technology in the context of a working commercial building, conducting a participatory learning experience in which planners, architects, builders, manufacturers and bankers are involved in the design and construction phase, and documenting the learning experience to provide a record and guide for others.

The Massachusetts Audubon Society has engaged Arthur D. Little, Inc., working with the architectural firm of Cambridge Seven Associates, to perform an initial planning study of the solar building. The study has the following elements:

1) Definition of the building requirements in terms of user needs and siting;

2) Preliminary architectural and engineering studies to establish the major elements of the solar space conditioning equipment and how these might be integrated into the building;

3) Preparation of cost estimates for the basic building and the incremental costs for the solar system;

4) Identification of participants who might bring products, services, or financial support to the design and construction phase.

A building configuration was selected which permits a reasonably large solar collector (3,500 square feet) facing south at an angle of 45° and which accommodates the MAS requirements for office space, a library and a lecture hall—a total floor space of approximately 8,000 square feet. A sketch showing the relationship of the proposed solar building to the current headquarter's building is presented in Figure 10.11. The section of the solar building is such that on the north side the roof height and exterior finish can be compatible with the adjacent existing building.

FIGURE 10.11: MAS PROPOSED SOLAR BUILDING AND SURROUNDINGS

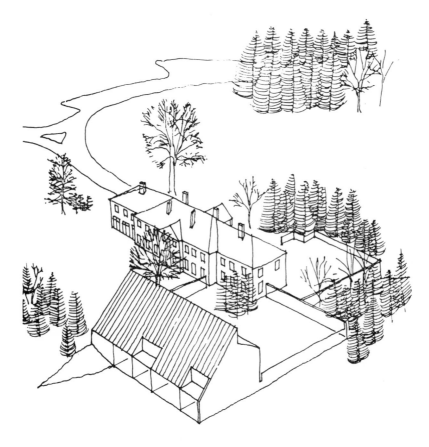

Source: PB-223 536, July 1973

Estimates of the heat loss indicate that depending on the choice of certain design options the heat demand will be in the range of 40,000 to 70,000 Btu/DD. It is estimated that a solar system using a two-pane 3,500 square foot collector should account for between 65 and 75% of the total seasonal heating load. Heat gain estimates indicate a design air conditioning load of about 15 tons, most of which is due to internal heat generation, (i.e., lighting and people). By making skillful use of natural ventilation it should be possible to minimize the periods of time during which complete air conditioning is required. Further, it appears that a 15 ton lithium bromide water cooled absorption machine driven by the solar collector should be able to supply a substantial portion of the air conditioning requirements.

Fort Belvoir, Va. Mech-Tech Building

The material in this section is from AD-A-002 576, *Solar Heating and Cooling of Buildings Study Conducted for Department of the Army, Volume I: Executive Summary and Implementation Plans,* prepared by General Electric, issued June 1974. Supplementary material is also available as AD-A-003 563. The purpose of this study was to select an existing building for a retrofit solar heating system and a new building to be constructed in the near future which could use a solar heating and cooling system. The latter system for the Fort Huachuca, Arizona classroom building, is discussed in the next section.

The characteristics of the Mech-Tech Building are summarized in Table 10.1. A heating requirements analysis of the building using a thermal model of the building, and hour-by-hour weather and solar data tapes for the Washington, D.C. area for 1963, a representative year, indicated a heating season requirement of 995×10^6 Btu.

TABLE 10.1: CHARACTERISTICS OF MECH-TECH BUILDING

- TWO STORY, FLAT ROOF, L-SHAPE
- BRICK AND CONCRETE BLOCK
- 15,000 SQUARE FEET FLOOR AREA
- 16 YEARS OLD
- 49 FAN COIL UNITS FOR SUMMER/WINTER OPERATION
- HEATING SYSTEM —
  STEAM TO HOT WATER CONVERTER FROM CENTRAL FACILITY
- COOLING SYSTEM —
  CHILLED WATER FROM ON-SITE CHILLER
- OFFICE AREAS, 60 PEOPLE
- USED 0800 TO 1700 HOURS, 5 DAYS A WEEK

Source: ADA-A-002 576, June 1974

The rooftop solar collector installation for providing a portion of the required heating is shown in Figure 10.12. A total of 130 collectors is used having a collection area of 3,055 square feet.

FIGURE 10.12: FORT BELVOIR MECH-TECH BUILDING

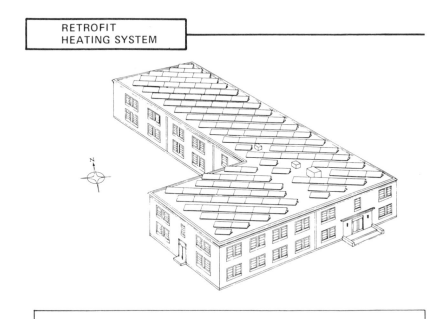

RETROFIT
HEATING SYSTEM

SYSTEM CHARACTERISTICS

SOLAR COLLECTORS

- PANEL AREA: 3055 SQUARE FEET
- PANEL SIZE: 38.5 X 96 INCHES
- NUMBER OF PANELS: 130
- ELEVATION ANGLE: 45 DEGREES
- AZIMUTH ANGLE: 11 DEGREES WEST OF SOUTH
- COLLECTOR TYPE: DOUBLE COVER WITH SELECTIVE ABSORBER COATING
- CIRCULATING FLUID: ETHYLENE GLYCOL-WATER MIXTURE

THERMAL STORAGE

- MEDIUM: WATER
- CAPACITY: 4000 GALLONS

ENERGY SAVED

- SPACE HEATING: 4000-5000 GALLONS FUEL OIL PER YEAR
- SUMMER DOMESTIC HOT WATER OPTION: 2000 GALLONS FUEL PER YEAR

Source: AD-A-002 576, June 1974

The collectors are 38.5 x 96 inches in size and are oriented at an elevation angle of 45° and an azimuth angle of 11° west of due south. This combination of angles maximizes the amount of solar energy collected. The flat plate collectors consist of a selectively coated absorber plate with integral fluid passages for heat transport and a double glazing of Lexan plastic windows, which entrap the incoming solar flux and minimize reradiation.

A schematic of the recommended solar energy system integrated with the existing heating system is shown in Figure 10.13. Several variations of this system were evaluated, with this arrangement determined to be the best choice considering performance and complexity.

FIGURE 10.13: SOLAR HEATING SYSTEM SCHEMATIC

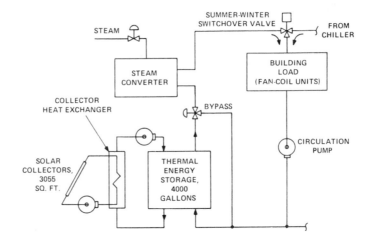

Source: AD-A-002 576, June 1974

Hot fluid (a glycol-water mixture) from the solar collectors is circulated through a heat exchanger whose secondary fluid (water) transports the energy to a 4,000-gallon thermal storage tank. Hot water from the tank is circulated through the building fan-coil units. Appropriate by-pass controls are used and auxiliary heat supplied from the existing steam converter when needed to provide circulating water at required temperature levels.

Hour-by-hour computer calculations were performed using the system described above together with the hour-by-hour heating requirements analysis described earlier. A computer summation of the delivered solar energy indicates that the solar system can supply 39% of the heating season requirements. The solar energy provides an annual savings of 4,000 to 5,000 gallons of fuel oil. This range in savings is due to the uncertainty in the efficiency of the existing central facilities steam distribution system in use at Fort Belvoir. Incorporation of energy conserving features, such as storm windows, nighttime temperature setback and reduced nighttime ventilation, reduces the total heating requirement and increases the fraction contributed by solar heating to 48%. Extending the system to provide domestic hot water to a nearby barracks complex during the summer could result in an additional savings of 2,000 gallons of fuel oil. The maximum number of solar collectors were installed on the roof considering the practical aspects of area occupied by existing equipment and the need for service walkways.

Implementation of this system is recommended. Except for the solar collectors, for which cost savings developments are continuing, the retrofit solar heating system can be implemented with commercially available material and equipment. The projected cost of implementation is $338,600.

Fort Huachuca, Ariz. Classroom Building

This information is from AD-A-002 576, *Solar Heating and Cooling of Buildings Study Conducted for Department of the Army, Volume I: Executive Summary and Implementation Plans,* prepared by General Electric, issued June 1974. Supplementary material is available as AD-A-002 563. The characteristics of the classroom building are summarized in Table 10.2.

TABLE 10.2: CHARACTERISTICS OF CLASSROOM BUILDING

- ONE STORY, FLAT ROOF
- 16,000 SQ. FT.
- REINFORCED CONCRETE SLUMP BLOCK
- ONLY 10% WINDOW AREA
- SPANISH MOTIF
- HOT WATER AND CHILLED WATER FROM CENTRAL FACILITIES PLANT
- ALL CLASSROOMS, EXCEPT FOR MECHANICAL ROOM AND SPACE FOR SMALL TRAINING COMPUTER
- USED 0800 TO 1700 HOURS, 5 DAYS A WEEK
- MAXIMUM OCCUPANCY OF 540 PEOPLE

Source: AD-A-002 576, June 1974

This building is part of a complex planned for the new academic service school at Fort Huachuca. Using defined occupancy schedules, equipment usage, building

characteristics, and heating and cooling requirements, analyses were performed
on a thermal model of the building and hourly weather and solar data for 1963
for the Phoenix, Arizona region. Modifications were introduced for adjusting
the data to Fort Huachuca conditions. The building was divided in four zones,
and the results produced hour by hour heating and cooling requirements for
each zone. Both rooftop and detached solar collection systems were studied.
The rooftop approach is recommended because of lower losses and is illustrated
in Figure 10.14.

FIGURE 10.14: FORT HUACHUCA CLASSROOM BUILDING

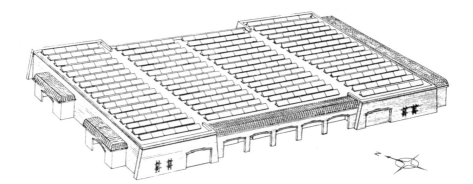

```
SYSTEM CHARACTERISTICS

SOLAR COLLECTORS

    • PANEL AREA:  7,040 SQUARE FEET
    • PANEL SIZE:  38.5 X 96 INCHES
    • NUMBER OF PANELS:  300
    • ELEVATION ANGLE:  25 DEGREES
    • AZIMUTH ANGLE:  27 DEGREES WEST OF SOUTH
    • COLLECTOR TYPE:  SINGLE COVER WITH SELECTIVE ABSORBER COATING
    • CIRCULATING FLUID:  ETHYLENE-GLYCOL-WATER MIXTURE

THERMAL STORAGE

    • MEDIUM:  WATER
    • CAPACITY:  12,000 GALLONS

COOLING EQUIPMENT

    • 100-TON ABSORPTION CHILLER (ARKLA)

ENERGY SAVED

    • HEATING AND COOLING – 2 MILLION CUBIC FEET OF GAS PER YEAR
```

Source: AD-A-002 576, June 1974

A total of 300 collectors is used having a collection area of 7,040 square feet. The collectors are 38.5 x 96 inches in size and are in-line with the roof edge at an azimuth of 27° west of south. The in-line arrangement was suitable because the performance of the collectors was insensitive over a wide azimuth range varying from due south. The system optimized for a low angle of 25° because of the dominating demand for cooling and the southern latitude of Fort Huachuca. The solar collectors are similar to those recommended for the Fort Belvoir, Va. building except that a single glazing is preferred over a double glazing. The single glazing unit is more cost effective in this operation.

A diagram of the recommended system concept is shown in Figure 10.15. The collected solar energy is stored as water sensible heat in a 12,000-gallon thermal storage tank. Heating is provided by circulating water from the thermal storage tank directly through coils in the building air duct system. Cooling is provided by circulating the hot water through a 100-ton absorption chiller which supplies chilled water to coils in the air duct system.

The results of parametric studies indicate that with 7,040 square feet of collectors the system can provide 98% of the heating requirements and 90% of the cooling requirements with an annual fuel saving of over 2 million cubic feet of gas (equivalent to 17,000 gallons of oil). These studies were performed on an hour-by-hour basis for the entire year since both heating and cooling is desired. The studies also considered the use of an economy cycle (using outside air directly for cooling) when feasible.

Incorporation of a solar heating and cooling system into this academic classroom building is recommended. Except for the solar collectors, for which cost savings developments are continuing, the system can be implemented with commercially available material and equipment. The projected incremental costs for implementation of the solar heating and cooling system are $465,000.

PPG Project

A privately financed project developed by Oliver Tyrone Corp., PPG Industries, Alcoa and Standard Oil Co. (Ohio) will include a 20' by 20' mockup, at PPG's Pittsburgh testing site, to test solar heat collection according to *Chemical and Engineering News,* September 2, 1974. This test is expected to be followed at some point by the construction of a 6- to 10-story building which will utilize solar heating and cooling.

Citicorp Center—New York City

MIT is studying the feasibility of placing a 20,000 square foot collector atop the 56-story Citicorp building under construction in New York City according to *Chemical and Engineering News,* February 10, 1975. The collected energy would be used to help operate dehumidifiers and heating, cooling and hot water systems.

FIGURE 10.15: SOLAR HEATING AND COOLING SYSTEM SCHEMATIC

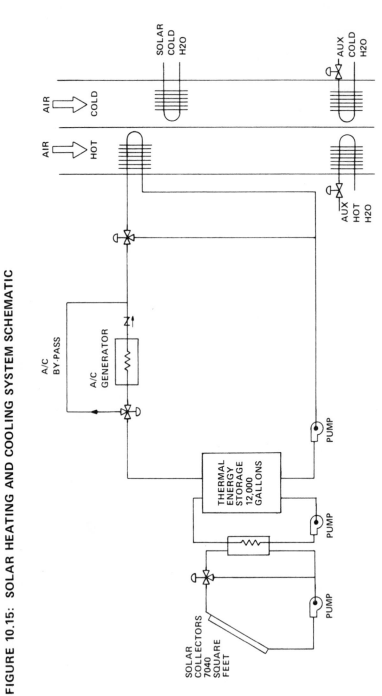

Source: AD-A-002 576, June 1974

Institute of Gas Technology Model

A climate control system designed for both heating and cooling using solar panels is being studied for production in 1977 by the Institute of Gas Technology, an affiliate of the Illinois Institute of Technology.

As reported in *Business Week,* May 18, 1974 tests on a full-size demonstration model indicate the system may be competitive costwise with conventional systems. IGT claims the model is 60% more efficient than existing gas-powered air conditioners. Air is the circulating medium in this system; there is no intermediary fluid.

General Services Administration Building—Manchester, N.H.

A feasibility study has been done for a solar heating and cooling system for the GSA's proposed energy conservation test building scheduled for completion in 1976 in Manchester, N.H. The seven-story 126,000 square foot structure will have 6,000 square feet of solar collectors on the roof to supply the heating and cooling needs of three floors.

The collector design recommended is a saw-tooth configuration which will reduce both wind loads and construction costs. Heat storage will be in three water tanks with a 30,000 gallon total capacity. The feasibility study was carried out by Dubin-Mindell-Bloome Associates, consulting engineers, and E.M. Barber, Yale University.

New York Botanical Gardens—Millbrook, N.Y.

A 30,000 square foot office and laboratory building having a 5,000 square foot collector is under construction at the Cary Arboretum in Millbrook, New York. The collector will be composed of seven elongated panels positioned on a horizontal roof. Heat pumps will also be employed as part of the system. The architect for the project is Malcolm Wells and the engineers are Dubin-Mindell-Bloome.

GENERAL ELECTRIC STUDY OF SHACOB RESEARCH

A summary of ongoing projects was included in PB-235 433, *Solar Heating and Cooling of Buildings, Phase 0, Feasibility and Planning Study, Final Report,* by General Electric Company prepared for the National Science Foundation and issued May 1974. This summary follows in Tables 10.3 through 10.8.

TABLE 10.3: SHACOB RESEARCH—SYSTEMS DESIGN AND PERFORMANCE

ORGANIZATION/ RESEARCHER	RESEARCH DESCRIPTION	RESEARCH OBJECTIVE	PRESENT STATUS	RESEARCHERS CONCLUSIONS
ARTHUR D. LITTLE, INC./ J. BURKE	SOLAR BUILDING PLANNING STUDY FOR MASSACHUSETTS AUDUBON SOCIETY	BUILDING OF A SOLAR-SPACE CONDITIONED BUILDING FOR OFFICE USE.	BUILDING IN SCHEMATIC DESIGN PHASE IN EARLY 1973	
CALIF. INST. OF TECH. & U.C.L.A./J. WEINGART & R. SHOEN	EXPLORATION OF THE POTENTIAL FOR COMMERCIALIZATION OF GAS SUPPLEMENTED SOLAR WATER HEATING SYSTEMS FOR APARTMENTS	USE OF GAS SUPPLEMENTED SOLAR WATER HEATING SYSTEMS IN SOUTHERN CALIF. APARTMENT UNITS	STUDY BEGUN UNDER NSF AUSPICES	
RICHARD RITTLEMAN, BUTLER, PENNA.	SOLAR HOUSE STUDIES	CONSTRUCTION OF DEMONSTRATION RESIDENCE IN WEST VIRGINIA	RESULTS OF PRELIMINARY STUDIES PUBLISHED	
SANDIA LABORATORIES/ R. STROMBERG	SOLAR COMMUNITY SYSTEM STUDIES INVOLVING CONCENTRATING COLLECTORS	DETERMINATION OF SYSTEM REQUIREMENTS FOR SOLAR COMMUNITY	STUDIES IN INITIAL STAGES	
S. MOUNTAIN DESIGN INC./ H. BARKMANN	PLANNED COMMUNITY INVOLVING SOLAR HEATED BUILDING	CONSTRUCTION OF SOLAR COMMUNITY	ARCHITECTURAL AND SOME COST STUDIES BEGUN, BUT MAJOR PROGRAM HAS FALLEN THROUGH	
UNIVERSITY OF PENN/ H. LORSCH & M. WOLF	COMPUTER STUDIES OF SOLAR HEATING, COOLING, AND PHOTOVOLTAIC SYSTEMS	SOLAR HEATED AND AC BUILDING	SOLAR HEATING SYSTEM DESIGNED AND ANALYZED. SOLAR AC SYSTEMS UNDER DEVELOPMENT	SOLAR HEATING READY FOR APPLICATION NOW: SOLAR AC 1-2 YEARS AWAY
UNIV. OF WISCONSIN/ J. DUFFIE	IDENTIFICATION AND OPTIMIZATION OF PRACTICAL SOLAR HEATING AND COOLING SYSTEMS USING SYSTEMIZATION	OPTIMIZE SYSTEMS FROM ECONOMIC POINT OF VIEW	MODELING OF PHYSICAL/THERMAL SYSTEMS ACCOMPLISHED IN EARLY 1973	

Source: PB-235 433, May 1974

TABLE 10.4: SHACOB RESEARCH—SOLAR COLLECTOR CHARACTERISTICS

ORGANIZATION/ RESEARCHER	RESEARCH DESCRIPTION	RESEARCH OBJECTIVE	PRESENT STATUS	RESEARCHERS CONCLUSIONS
UNIV. of DEL/ K. Boer	COMBINED THERMAL-PHOTOVOLTAIC COLLECTOR	PROVIDE THERMAL AND ELECTRICAL ENERGY FOR DWELLING CONSUMPTION FROM SOLAR ENERGY. (SUPPLY UP TO 80% OF DEMAND)	TEST UNITS INSTALLED ON SOLAR HOUSE, DATA COLLECTION BEGUN	
UNIV. of MELBOURNE AUSTRALIA/W. CHARTERS	THEORETICAL EXPERIMENT STUDIES OF HEAT TRANSFER PROCESSES IN VARIOUS COLLECTOR CONFIGURATIONS	PROVIDE HEAT TRANSFER INFORMATION FOR USE IN COLLECTOR DESIGN	STUDIES OF TILTED HONEY-COMB V-CHANNEL COLLECTORS PUBLISHED. ALSO MULTIPLE COVER STUDIES INVOLVING INNER PLASTIC COVER	USE OF HONEYCOMB STRUCTURE ON TILTED FLAT-PLATE COLLECTO ON TILTED FLAT-PLATE COLLECTO NOT WARRANTED. FLAT-PLATE SUPERIOR TO V-CHANNEL DUE TO CONVENTION LOSSES. COST ADVANTAGES OF 2nd COVER NEED EXPLORING.
UNIV. of PENN/ H. Lorsch	ANALYTICAL AND EXPERIM STUDIES OF FLAT-PLATE COLLECTORS	DEVELOPMENT OF FLAT-PLATE COLLECTORS	PROTOTYPE SIZE COLLECTOR INSTALLED IN TEST FACILITY	USE OF SELECTIVE COATING NOT COST EFFECTIVE. OPTIMUM SPACING FOR 2 PANE COLLECTORS IS $\frac{3}{8}" - \frac{1}{2}"$
AMERICAN CYANAMID/	CADMIUM STANNATE COATINGS TO ENHANCE GREEN HOUSE EFFECT PROPERTIES OF GLASS	INCREASE EFFICIENCY OF SOLAR ENERGY CAPTURE	DEVELOPMENTAL; TECH. FEASIBILITY NOT ESTABLISHED. ECON. FEASIBILITY NOT ESTABLISHED	

(continued)

TABLE 10.4: (continued)

ORGANIZATION/ RESEARCHER	RESEARCH DESCRIPTION	RESEARCH OBJECTIVE	PRESENT STATUS	RESEARCHERS CONCLUSIONS
BATTELLE-COLUMBUS/ J. A. EIBLING	MULTILAYER SELECTIVE ABSORBER COATINGS, INCLUDING Ni/SiO_2	REDUCE COST TO $0.20-$0.30 PER FT^2		
UNIV. OF ARIZONA/ B.O. SERAPHON AND V.A. WELLS	ABSORBER COATINGS DEPOSITED BY CHEMICAL VAPOR TRANSPORT TECHNIQUES. PRIMARILY Ag-Si-Si Nu - Si O	$\alpha/\epsilon > 20$ LOW COST MANUFACTURE LONG LIFE - LOW DEGRADATION	$\alpha/\epsilon \sim 10 - 12$ ECON FEASIBILITY NOT ESTABLISHED; LIFE NOT TESTED	THIS TYPE ABSORBER SURFACE LESS SUSCEPTIBLE TO DEGRADATION, LONGER LIFE REDUCES COST
UNIV. of CALIFORNIA M. E. MERRIAM (BERKELEY)	1. SELECTIVE ABSORBER SURFACES OF SEMICONDUCTOR AND GEOMETRIC TRAPPING TYPES	LOW COST COATINGS TO IMPROVE PERFORMANCE	PROPOSED 1973	
	2. PLASTIC AND PLASTIC-GLASS HYBRIDE AS GLASS SUBSTITUTES	LOW COST COVER PLATES WITH HIGH SOLAR XMITTANCE, LOW IR XMITTANCE	PROPOSED 1973	
	3. A R COATINGS FOR GLASS AND POSSIBLY PLASTICS	LOWER COST COATINGS	PROPOSED 1973	
TEXAS INSTRUMENTS/ T. SANTALA & W. PAYNTON	ABSORPTION SURFACE USING Fe_2Al_5 (multiple SCATTER ABSORPTION)	LOW COST, HIGH EFFICIENCY SOLAR COLLECTOR PANEL	$\alpha_s = 0.955$ $\epsilon_{IR} \sim 0.63 - 0.78$ DEVELOPMENTAL	INTERMETALLICS (Fe_2Al_5) HAVE EXCELLENT POTENTIAL

Source: PB-235 433, May 1974

TABLE 10.5: SHACOB RESEARCH—SOLAR ENERGY STORAGE AND TRANSFER

ORGANIZATION/ RESEARCHER	RESEARCH DESCRIPTION	RESEARCH OBJECTIVE	PRESENT STATUS	RESEARCHERS CONCLUSIONS
UNIV. OF DELAWARE / M. TELKES	HEAT STORAGE USING PHASE CHANGE MATERIALS	COMPACT, EFFICIENT, LOW COST STORAGE COMPONENTS	SALT HYDRATE TYPE RESERVOIRS INSTALLED IN EXPERIMENTAL SOLAR HOUSE	COST OF MATERIAL NEAR $0.30 / LB 10 YEAR LIFETIME
UNIV. OF PENNSYLVANIA	EXPERIMENTAL INVESTIGATION OF A VARIETY OF PHASE CHANGE MATERIALS SUITABLE FOR SPACE AND WATER HEATING AS WELL AS ABSORPTION AC APPLICATIONS	WELL BEHAVED, LONG LIFE HEAT STORAGE MATERIALS FOR BROAD SPECTRUM OF SOLAR APPLICATIONS	A NUMBER OF MATERIALS STUDIED, SOME FELT TO BE IMMEDIATELY APPLICABLE TO SHACOB	ENGINEERING SPECIFICATION OF OPERATIONAL PARAMETERS SUCH AS TEMP. NEEDED. USABLE MATERIALS AVAILABLE FOR SOME APPLICATIONS

TABLE 10.6: SHACOB RESEARCH—SOLAR HEATING TECHNIQUES

ORGANIZATION/ RESEARCHER	RESEARCH DESCRIPTION	RESEARCH OBJECTIVE	PRESENT STATUS	RESEARCHERS CONCLUSIONS
G. LOF, DENVER, COLO.	SOLAR AIR HEATERS	LOW COST HIGH EFF. SOLAR HEATERS	LIMITED DATA PUBLISHED	AIR HEATERS SHOULD NOT BE OVERLOOKED IN SELECTING SPACE HEATING SYSTEMS

Source: PB-235 433, May 1974

TABLE 10.7: SHACOB RESEARCH—SOLAR COOLING TECHNIQUES

ORGANIZATION/ RESEARCHER	RESEARCH DESCRIPTION	RESEARCH OBJECTIVE	PRESENT STATUS	RESEARCHERS CONCLUSIONS
ALLIED CHEMICAL CORP.	DEVELOPMENT OF ORGANIC REFRIGERANTS	CUSTOM MADE REFRIGERANTS		
ARKLA SERVEL CORP.	DEVELOPMENT OF ABSORPTION AIR CONDITIONERS WHICH ARE SOLAR COMPATIBLE		ACTIVELY WORKING ON CONVERTING SOME OLDER UNITS TO HOT WATER FIRED FOR SOLAR USE	
COLORADO STATE UNIV./ G. LÖF	PROPOSED (1973) INSTALLATION OF FULL SCALE SOLAR HEATING AND COOLING SYSTEM IN EXPERIMENTAL HOUSE USING 3-TON ARKLA-SERVEL AIR CONDITIONER	DEMONSTRATE FEASIBILITY OF SOLAR COOLED BUILDING, VERIFY MODELS, TEST FACILITY FOR NEW AND IMPROVED DESIGNS	PROP 1973, WORK BEGUN	
G. LÖF, DENVER, COLO.	SOLAR COOLING DESIGN AND COST STUDY	DETERMINE SYSTEM COST PER MILLION BTU OF SOLAR ENERGY DELIVERED AND OPTIMIZE SYSTEM	PUBLISHED DATA FOR OPTIMUM SYSTEM AT 8 LOCATIONS IN THE USA	SOLAR COOLING CAN SUBSTANTIALLY REDUCE SOLAR ENERGY SUPPLY COSTS. COMBINED SYSTEM APPROACHING COMPETITIVE POSITION WITH DOMESTIC FUEL
ENERGY TECHNOLOGY, INC. UNIV. OF PENN	AC UNIT FEATURING COMPRESSOR DRIVEN BY STEAM POWERED TURBINE, USING SOLAR HEAT	VAPOR-COMPRESSION AC SOLAR POWERED	HIGH PRESSURE TURBINE ASSEMBLED LOOKING FOR SPONOSR FOR LOW PRESSURE SOLAR POWERED TURBINE	

(continued)

TABLE 10.7: (continued)

ORGANIZATION/ RESEARCHER	RESEARCH DESCRIPTION	RESEARCH OBJECTIVE	PRESENT STATUS	RESEARCHERS CONCLUSIONS
THERMO ELECTRON CORP./ W. TEAGEN	VAPOR COMPRESSION AC SYSTEM DRIVEN BY ORGANIC RANKINE ENGINE	FLEXIBLE, COST COMPETITIVE SYSTEM WITH EFFICIENT OPERATION IN 160°-280°F TEMP. RANGE	CALCULATED PERFORMANCE CURVES PUBLISHED. EQUIPMENT STATUS SAME AS ABOVE	EFFICIENCIES OF 5%-17% CAN BE ACHIEVED AT FLAT PLATE COLLECTOR TEMPERATURES
UNIV. OF FLORIDA/ E. FARBER	ABSORPTION AC UNIT USING NH₃-H₂O SYSTEM	STUDIES OF SOLAR AC PERFORMANCE	5 TON UNIT IN OPERATION	
UNIV. OF MARYLAND/ R. ALLEN	EFFECT OF SYSTEMS OPTIONS AND PROCESS FACTORS ON PERFORMANCE AND OPTIMIZATION OF SOLAR POWERED ABSORPT. AC UNITS	PREDICT RELATIVE COSTS AND PERFORMANCE OF CANDIDATE SYSTEMS	PROP 1973	

Source: PB-235 433, May 1974

TABLE 10.8: SHACOB RESEARCH—EXPERIMENTAL STUDIES

ORGANIZATION/ RESEARCHER	RESEARCH DESCRIPTION	RESEARCH OBJECTIVE	PRESENT STATUS	RESEARCHERS CONCLUSIONS
CENTRE NATIONAL DE LA RECHERCHE SCIENTIFIQUE, FRANCE	SOLAR HEATED BUILDINGS USING WALLS AS COLLECTOR AND AIR CIRCULATION	PERFORMANCE DATA ON SOLAR HEATED BUILDINGS	BUILDINGS IN EXISTENCE SEVERAL YEARS	
DUBIN-MINDEL-BLOOME ASSOC.	DESIGNED A NUMBER OF SOLAR HEATED BUILDINGS	PERFORMANCE DATA ON SOLAR HEATED BUILDINGS		
PENNSYLVANIA POWER AND LIGHT	CONSTRUCTION OF ENERGY CONSERVING HOUSE	SAME AS ABOVE		
SKY THERM/ H. HAY	CONSTRUCTION OF SOLAR HOUSE FOR EVALUATION OF NATURAL RADIATION FLUX HEATING AND COOLING	SAME AS ABOVE	BUILDING BEING EVALUATED IN CONJUNCTION WITH CALIF. POLYTECHNIC UNIV.	
UNIV. OF DELAWARE/ K. BOER	CONSTRUCTION OF EXPERIMENTAL SOLAR HOUSE USING THERMAL PHOTOVOLTAIC COLLEC-TOR	PERFORMANCE DATA ON SOLAR HEATED AND COOLED BUILDING	BUILDING CONSTRUCTED, TESTS UNDERWAY	

(continued)

TABLE 10.8: (continued)

ORGANIZATION/ RESEARCHER	RESEARCH DESCRIPTION	RESEARCH OBJECTIVE	PRESENT STATUS	RESEARCHERS CONCLUSIONS
CSIRO, AUSTRALIA/ P.S. SEANES	DETERMINATION OF SEQUENCES OF CLIMATE DATA USEFUL IN HEATING PLANT DESIGN AND TEMP. EXTREME DETERMINATION	FAST, ACCURATE THERMAL DESIGN EVALUATIONS	PRACTICAL METHODS OF DERIVING SEQUENCES OF DATA FOR DESIGN USE DISCUSSED IN PUBLICATION	DESIGN DATA MUST BE ASSOCIATED WITH BUILDING TYPE
HITTMAN ASSOC. INC.	ENERGY CONSUMPTION IN BUILDINGS (GENERAL)	RECOMMENDATIONS FOR ENERGY CONSERVATION DETERMINATION OF HEAT PRODUCING SOURCES IN BUILDING	PUBLISHED RESULTS ON STUDIES SPONSORED BY HUD	
NBS / J. HILL / T. KUSUDA	COMPUTER PROGRAM SIMULATING BUILDING ENVIRONMENT INTERACTION	DETERMINE EFFECT OF BUILDING DESIGN PARAMETERS ON BUILDING THERMAL RESPONSE	PROGRAM DEVELOPED AND IN USE.	PROPER BUILDING DESIGN RESULTING FROM THIS TYPE PROGRAM COULD MAKE SOLAR SYSTEMS MORE COMPETITIVE
OHIO STATE UNIV. / C. SEPSY	DEVELOPMENT OF COMPUTER PROGRAMS TO SIMULATE BUILDING ENVIRONMENTAL CONTROL SYSTEMS	IMPROVED LOAD CALCULATION AND SYSTEM SIMULATION PROCEDURES	PROGRAMS UNDER DEVELOPMENT ON NSF-RANN PROGRAM	
PRINCETON UNIV. / R. GROT	STUDY OF ENERGY CONSUMPTION PATTERNS IN A PLANNED COMMUNITY (TWIN RIVERS, N.J.)	ACCUMULATION AND ANALYSIS OF DATA TO CORRELATE ENERGY CONSUMPTION WITH HOUSE TYPE, WEATHER, RESIDENT CHARACTERISTICS, ETC.	STUDY UNDERWAY, OVER 1500 DWELLING UNITS OCCUPIED IN COMMUNITY.	CORRELATION FACTORS LOW IN INITIAL STUDIES

Source: PB-235 433, May 1974

Commercially Available Solar Hardware

In response to inquiries the following information was supplied, by the companies noted, on several available solar components.

AAI COST EFFECTIVE SOLAR COLLECTOR

AAI Corporation has designed and built the Timonium Elementary School building's solar heating system as described above. The system uses AAI's double-glass honeycomb collector which is efficient and producible at low cost.

The intention had been to produce this unique collector for the market however, through continued research and development work, AAI has now developed a new type collector that they feel is a major break-through in establishing a cost-competitive system for heating and cooling of buildings. This unit, which they call a Roof-Top Concentrator, is composed of a movable collector located above and parallel to a curved mirror surface which can serve as the building's roof. Roof-Top Concentrators collect more than twice the energy per year than do the best flat plate collectors and they do this at less than half the cost of flat plates.

AAI investigations and those of other major organizations in the solar energy field have shown that flat plate collectors, when priced in high quantity production, will produce energy at a cost of over $7 per million Btu. This compares with the cost of energy production by oil systems of $2 to $4 per million Btu.

The development program for the Roof-Top Concentrator shows it will deliver solar energy at a cost under $2 per million Btu. Test models in the 100 sq ft size have shown efficiencies greater than originally calculated. Its use is foreseen primarily with absorption-type air conditioning in the summer (water temperature of 220°F), and hot water heating in the winter.

AAI is now in the process of carrying out a full-scale experiment to prove this equipment. The next objective is to establish a full-scale demonstration system with a unit of 5 to 25 thousand sq ft for a public facility. The intention is to produce and market Roof-Top Concentrators after demonstrations are completed. Plans are to market the collector, the control mechanism and the pumping system while specifying the means of construction of the concentrator which would become a part of the roof on a new building.

PPG BASELINE SOLAR COLLECTOR

PPG's Baseline Solar Collector design is based on 30 years' research, production and field experience with hermetically sealed insulating glass units. Although a new product, preliminary testing and development have been completed. PPG Baseline Solar Collectors have successfully endured aging tests at 300°F. Various solar installations in the United States are now using PPG's Baseline Solar Collector. The PPG Baseline Solar Collector is shown in Figure 11.1.

PPG Baseline Solar Collectors can be used in solar-powered space heating and hot water systems, where a fluid is used to transport heat from the collector. PPG's Baseline Solar Collector is a flat plate-type collector, constructed with the following materials.

(1) Unit size: 34³⁄₁₆" x 76³⁄₁₆" x 1⁵⁄₁₆".

(2) Cover plates: two pieces of ⅛" Herculite tempered glass.

(3) Absorbing surface: PPG's Duracron Super 600 L/G (UC 40437) flat black coating.

(4) Collector plate: Roll Bond 1100 aluminum solar absorber.

(5) Insulating backing: 2½" backing of fiber glass.

(6) Metal pan insulation housing available.

(7) Edge retaining system: metal channel with desiccant-type spacers.

Glass Cover Plates: PPG's Baseline Solar Collector utilizes ⅛" Herculite fully tempered clear float glass because it provides high solar transmittance and is opaque in the reradiating infrared spectrum, i.e., beyond 2.8 microns.

Thermal Properties: Two cover plates are provided to reduce upward heat losses. Glass properties are as follows.

Emissivity at 0°–200°F	0.9
Specific heat at 212°F (32°–212°F)	0.205
Thermal stress endurance (degrees differential)	450°F

(continued)

Maximum working temperature (over entire glass area)
   Long exposure (greater than 100 hr)   450°-500°F
   Short exposure (less than 10 hr)   500°-550°F

FIGURE 11.1: PPG BASELINE SOLAR COLLECTOR

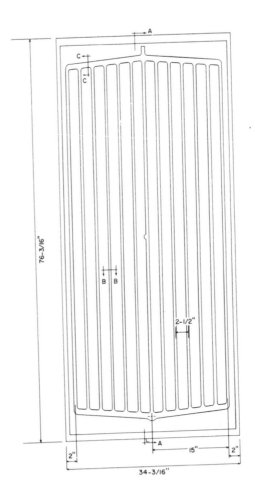

(continued)

FIGURE 11.1: (continued)

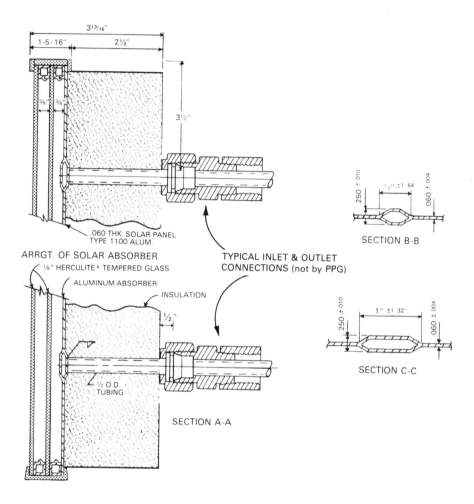

Source: PPG Industries, Pittsburgh, Pa.

Wind Load: The 34³⁄₁₆" x 76³⁄₁₆" x ⅛" Herculite insulating unit for the PPG Baseline Collector can withstand a one-minute, fastest mile uniform wind load of 225 psf with a probability of breakage equal to 8 lights per 1,000.

Snow Load: The 34³⁄₁₆" x 76³⁄₁₆" x ⅛" Herculite insulating unit for the PPG Baseline Collector can withstand a long-term (more than one hour) uniform snow load of 170 psf, with a probability of breakage equal to 8 lights per 1,000.

Hail: PPG $\frac{1}{8}$" Herculite glass has been shown to be effective against normal hailstone impact.

Absorbing Surface: PPG's Baseline Solar Collector utilizes a flat black coating for the collector plate. The coating is PPG's Duracron Super 600 L/G (UC 40437) product. The Duracron coating has a solar radiation absorbance, α, of 0.95 and an infrared emittance, ϵ, of 0.95 ($\alpha/\epsilon = 1.0$).

Collector Absorber Plates: PPG's Baseline Solar Collector utilizes a Roll Bond 0.060 aluminum (1100 type) collector absorber plate.

Insulating Backing: PPG's Baseline Collector utilizes fiber glass insulation backing.

Edge Retaining System: PPG's Baseline Solar Collector utilizes a special edge retaining system where the $\frac{1}{8}$" Herculite tempered glass plates are separated by a spacer. These spacers contain a desiccant which reduces condensation on the glass plates. This special design eliminates the effect of pressure build up with temperature changes. The edge attachment system for the unit consists of a metal channel with high temperature resisting sealants that have the capability to withstand high temperature no load conditions.

The price of the collection unit as of February, 1975 is $200, including boxing, FOB Ford City, Pa.

REVERE SOLAR ENERGY COLLECTOR

The Revere Solar Energy Collector is efficient, easy to install, long lasting, maintenance free and reasonable in cost. It can heat a fluid more than $100°F$ above the surrounding air temperature and is practical for a variety of uses, from heating a swimming pool to providing hot water, heat and air conditioning.

The Revere Solar Energy Collector is an extension of the established Revere System of Laminated Panel Construction. In the Revere panel system, copper sheet is laminated to plywood to make a copper composite building panel. Joints between the copper laminated panels are sealed with special roll-formed copper joint members and high grade sealing tape.

To convert an installation of the Revere laminated panel system to a solar collector, rectangular copper tubes are secured to standard 2 feet x 8 feet panels with copper clips. A special high conductivity adhesive is applied between tube and panel to insure a good heat transfer characteristic. Tube spacings may vary from 4" to 10" on center, depending upon efficiency requirements. A specially constructed copper batten supports one or two layers of glass or other transparent material.

Revere has also developed simple adapter fittings to facilitate the connection of the rectangular tubes to conventional round copper water tubes. An installation made in this manner can function as a combination roof and solar collector in new buildings. Preassembled modular units in 3 feet x 7 feet sizes are also available for use on existing structures.

Figure 11.2 shows the Revere Copper Laminated Panel System and its adaptation as a solar collector. The Revere Solar Energy Collector may be installed as a modular unit for existing structures, or it may serve a dual function as both roof and collector in new buildings.

Operation of the Revere Solar Energy Collector is simple. A blackened copper surface absorbs radiant heat and transfers it to a fluid circulated through tubes which are fastened securely to the absorber surface. The collector plate is usually covered with glass, or other transparent material, which admits radiant energy from the sun, but traps any energy which is reradiated from the warm surface.

Applications for solar energy collectors include heating domestic water, buildings and swimming pools, and for furnishing heat to operate absorption type air conditioning units. Solar energy collectors also have potential uses in commercial laundries, drying of agricultural products and other industrial applications. Each of these calls for the solar energy collector system to perform somewhat differently. An installation in which domestic water is heated does not usually require water temperatures much higher than 140°F. However, an application for heating a building might require water temperatures up to 170°F.

Table 11.1 summarizes design and component variables for Revere Solar Energy Collectors for the applications mentioned above.

TABLE 11.1: PERFORMANCE DATA

	HEATING			AIR CONDITIONING
	Swimming Pools	Domestic Water	Building	
Usual Fluid Temperature Range (°F)	70-90	100-140	120-200	160-230
Maximum Design Temperature of Fluid Above Ambient (°F)	25-30	Summer 60-80 Winter 80-110	Fall/Spring 80-120 Winter 100-150	90-140
Number of Tubes Per 2' Wide Panel	2-3	Summer 3-4 Winter 3-4	Fall/Spring 3-4 Winter 3-5	3-5
Usual Number of Transparent Covers	0	Summer 1-2 Winter 1-2	Fall/Spring 2-3 Winter 2-3	2-3

Note: Insulation for each of the above applications will vary according to geographical location and type of insulation used.

Source: Revere Copper and Brass, Inc., Rome, N.Y.

FIGURE 11.2: REVERE COMBINATION LAMINATED PANEL ROOF AND SOLAR COLLECTOR

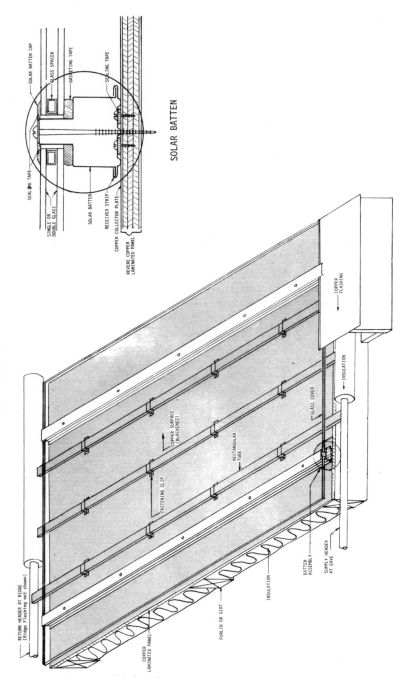

Source: Revere Copper and Brass, Inc., Rome, N.Y.

Revere Laminated Building Panels and Solar Energy Collectors are available in a 2 feet x 8 feet size and can function as both solar energy collector and roof or sidewall. Other sizes are available on special order. On these units the battens, tubes, cover pieces, receiver strips, headers, clips, conductive adhesive, sealing tapes and necessary adapter fittings are supplied by Revere.

To eliminate breakage and minimize shipping costs, glass for the solar collector may be procured locally. Glass with low iron content is best for transmitting the maximum amount of the sun's radiant energy. Transmittance values for a single sheet of glass will vary with the thickness, but should range from 83 to 90%. Transmittance values for two sheets of glass should range between 70 and 81%, but may be slightly lower for some two thickness glass assemblies.

The standard coating used on the Revere Solar Energy Collector is a high-quality, heat-resistant, flat black paint. With proper application this provides a surface which will absorb 95% of the solar energy which touches it.

The collector unit illustrated in Figure 11.3 usually will find its best application and advantage in new construction. The modular unit shown in Figure 11.3 is readily adaptable to either new or existing construction.

In the modular unit the spacing of tubes on the collector plate is adaptable to different design requirements, although 6" centers seems a common and effective spacing. Tube layout and header configurations can be adapted as required for grid, grid-sinuous or sinuous flow conditions. At both ends of the collector unit the ½" x 1" rectangular tubes are brazed to conventional ¾" round Type M copper water tube headers. The header ends project through the aluminum housing of the unit for joining by conventional soldering to inlet and outlet piping. Header ends can project through either the side or the bottom of the housing as desired to facilitate connections to the outside piping.

Insulation housing under the collector plate is 3½" fiberglass and around the perimeter of the housing is 1" styrofoam. Wood liners around the perimeter of the housing are optionally available to facilitate installation and securement to existing roof construction, angle iron framing, etc. Transparent coverings can be supplied in different numbers of thicknesses or layers depending on the application or type of heating to be provided. With a double glass cover the unit weighs approximately 8½ psf. The basic size of the modular unit is 36" x 78" x 5". The broad capabilities of design adaptability of this collector system to different heating or output requirements is a feature not available in other solar collector products and is an advantage of this system.

The cost of the collector unit for new construction as of February, 1975, is $5.90/ft^2 and the cost of the modular unit is $7.80/ft^2, both FOB Rome, N.Y.

FIGURE 11.3: REVERE MODULAR RETROFIT COLLECTOR

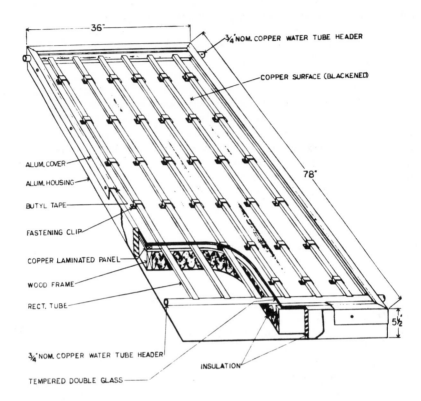

Source: Revere Copper and Brass, Inc., Rome, N.Y.

SUNWORKS FLAT PLATE COLLECTOR

Sunworks, Inc. is manufacturing a flat plate solar heat collector which is efficient, durable, and versatile, and it can be used for any application which requires low-grade thermal energy.

Heat is captured when solar radiation penetrates the collector's high transmittance glass cover and strikes the selectively blackened copper absorber, causing it to heat up. This heat warms the liquid in tubes imbedded in the absorber and is carried away from the absorber by the liquid to another part of the system, usually a heat storage container. The heat is then applied to the desired purpose—space heating, air conditioning (via a liquid absorption chiller), domestic water heating, various industrial processes, or whatever.

The Sunworks flat plate collector is available in two versions: a flush-mounted

module which is designed to become a part of the exterior membrane of the
building to which it is attached and a surface-mounted module which can be
attached to any surface or specially constructed framework with no need for
further weatherproofing.

Flush-Mounted Module

This module is designed primarily for integration with the roof or wall of a
building. It is designed to fit between joists or rafters spaced 24" on center.
It is available in 94" lengths with shorter lengths available at additional cost.
When installed, this module can become the exterior skin of the roof or wall.
Components for flashing the module are available. The modules can be mounted
side by side across the roof or wall and/or end to end going up the roof or wall.
Rafters or joists must be at least 5½" thick. A 2" x 6" (1½" x 5½") is the
smallest structural member that can be used. The flush-mounted module is
shown in Figure 11.4.

FIGURE 11.4: FLUSH-MOUNTED MODULE

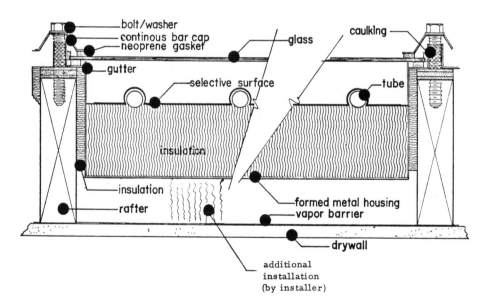

Source: Sunworks, Inc., Guilford, Conn.

Technical data on materials
    cover: single glass, 3/16", tempered, edges swiped
                  92% solar transmittance
    absorber container: galvanized sheet metal, 24 ga.
                  (other metals available on request)   (continued)

air space between cover and absorber: 1/2" - 3/4"

absorber:

        -copper sheet: 0.010" thick (7 oz.)

        -selective black: minimum absorptivity 0.90;
                          maximum emissivity 0.12
                          manufactured by Enthone, Inc.,
                          durable to 400° F.

        -copper tubes: $\frac{1}{4}$" $\emptyset$ (.375" O.D.) M-type

        -tube spacing: 6" on center

        -tube pattern: grid

        -manifolds: 3/4" $\emptyset$ M-type copper

        -tube connections to manifold: silver solder

        -bond between tube and sheet: solder

        -connection to external piping: 3/4" $\emptyset$
                          M-type copper, extending 2" beyond
                          collector ends; supply, top right;
                          return, bottom left (when viewed
                          from the top)

        -manifold/tubes pressure-tested to 15 atm.

insulation behind absorber: 3" thick fiberglass, 1.5 lb./ft.^{3}

gasketing material: neoprene

weatherproofing:

        -between collector mullions, furnished with
                collectors: 26 ga. galvanized steel

        -side roof mullions, furnished as requested:
                26 ga. galvanized steel

        -top flashing caps, furnished as requested:
                .020" terne sheet

<u>Dimensions of the flush-mounted module</u>

        -portion between rafters: 22" wide x 94" long

        -outside dimensions overall: 23-5/8" wide x 94" long

        -absorber surface: 21" wide x 90" long

        -ratio of usable absorber area to total surface
        covered: 0.86

        -distance module extends below uppermost surface of
        rafter: 4"

        -distance mullions extend above rafter: $1\frac{1}{2}$"

<u>Method of anchoring to rafters or joists</u>

        -$\frac{1}{4}$" $\emptyset$ - 3" lag bolts with gaskets, passed through
        mullions into rafters, furnished by installer

<u>Weight per module:</u> 70 pounds

<u>Recommended flow rate through the collector:</u>

      - 1 gph/sq. ft. of collector (F_r = .9)

      - flow resistance at this rate is negligible

<u>Collector Coolant</u>

    The coolant can be inhibited alcohol-water mixtures such as standard automobile anti-freeze made by Union Carbide or duPont. In areas where regular tap water is used as a coolant, it is important that the pH be controlled between 6 and 8. These collectors can be used with other coolants but the user must contact the manufacturer for approval of specific liquids.

Surface-Mounted Module

This module is intended for installation on any structurally sound surface or specially constructed framework. It can be mounted on a surface, such as an existing roof, and connected end to end or side by side. It is shown in Figure 11.5.

FIGURE 11.5: SURFACE-MOUNTED MODULE

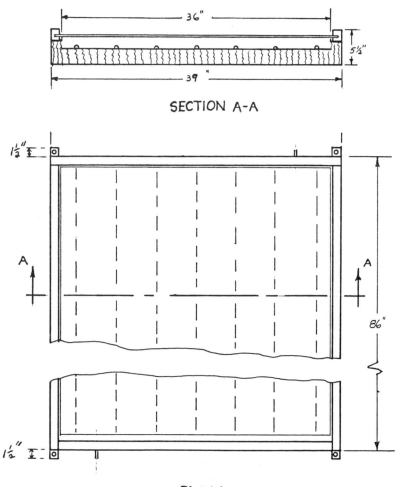

SECTION A-A

PLAN

Source: Sunworks, Inc., Guilford, Conn.

Technical data on materials

    cover: single glass, 3/16" thick, tempered, edges swiped
              92% solar transmittance

    absorber container: aluminum sheet, .04" and aluminum
                             extrusions

    air space between cover and absorber: 1/2" - 3/4"

    absorber:

            -copper sheet: 0.010" thick (7 oz.)
            -selective black: minimum absorptivity 0.90;
                      maximum emissivity 0.12
                      manufactured by Enthone, Inc.,
                      durable to 400° F.
            -copper tubes: $\frac{1}{4}$" ∅ (.375" O.D.) M-type
            -tube spacing: 6" on center
            -tube pattern: grid
            -manifolds: 1" ∅ M-type copper
            -tube connections to manifold: silver solder
            -bond between tube and sheet: solder
            -connection to external piping, 1" ∅
                      M-type copper, extending 2" beyond
                      collector ends; supply, top right;
                      return, bottom left (when viewed
                      from the top)
            -manifold/tubes pressure-tested to 15 atm.

    insulation behind absorber: 2" thick fiberglass,
                  1.5 lb/ft.^{3}

    gasketing material: glass: aluminum extrusion
                    all other: roofing sealant

    weatherproofing: this module can be placed out in the weather
                  without need for further weatherproofing

Dimensions of the surface-mounted module

            -outside dimensions overall: 39" wide by 86" long
            -absorber surface: 36" wide by 83" long
            -ratio of usable absorber area to total surface
            covered: 0.89
            -thickness of module: $5\frac{1}{2}$"

Method of anchoring

          -4 anchor tabs, with bolt holes, are fastened
          to each corner of the module

Weight per module: 120 pounds

Recommended flow rate through the collector:

       - 1 gph/sq. ft. of collector (F_r = .9)
       - flow resistance at this rate is negligible

Collector Coolant.

    The coolant can be inhibited alcohol-water mixtures such as standard automobile anti-freeze made by Union Carbide or duPont. In areas where regular tap water is used as a coolant, it is important that the pH be controlled between 6 and 8. These collectors can be used with other coolants but the user must contact the manufacturer for approval of specific liquids.

The cost of a single surface-mounted unit, when ordered in quantities of less than 10, as of January, 1975 is $246.00.

TRANTER SOLAR COLLECTOR PLATE

Tranter, Inc. is now producing a highly efficient flat plate solar collector, having about 90% internally wetted surfaces. A continuous welding process controls the internal water flow pattern to allow for four passes, two passes or full flooded designs.

General Specifications:

Allowable operating pressure: up to 75 psig

Thickness: approximately 22 gauge = approximately 0.029"

Material: carbon steel or stainless steel

Weight: approximately 55 lb for 2' x 10' size

Finish: supplied as welded with paint to be applied by others

Pressure drop: usually less than 5 psi for usual low flow rates in this service

Carbon steel is generally considered suitable for closed recirculating systems using glycol or treated water

Stainless steel is ideal for swimming pool water and for once through systems such as heating tap water

FIGURE 11.6: TRANTER SOLAR COLLECTOR PLATE MODELS

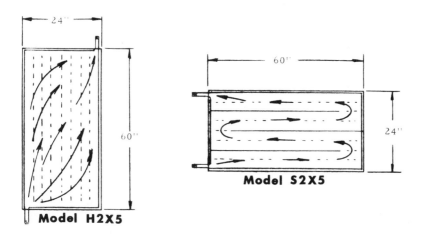

Model H2X5

Model S2X5

(continued)

FIGURE 11.6: (continued)

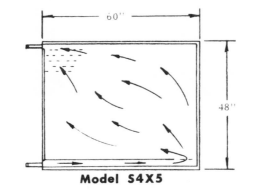

Model S4X5

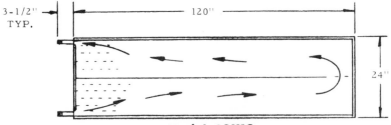

Model S2X10

MODEL NUMBER	AREA SQ. FT.	APPROX. WT. LBS.
H2X5	10	27
S2X5	10	27
S4X5	20	54
S2X10	20	54

Source: Tranter, Inc., Lansing, Mich.

All of the models are supplied in approximately 22 gauge carbon steel, leak tested at 125 psig, ready for painting and installation. Fittings are ¾" diameter threaded pipes. Other circuitry and sizes are available. Sizes may be reduced without price change.

Water flow rate for 10 sq ft models at 10 lb/hr/sq ft = 0.2 gpm and 0.4 gpm for the 20 sq ft size models. Pressure drop for water at up to 1 gpm will be not over 1 psi for any model.

NASA Lewis in Cleveland recently completed testing the Econocoil Solar Collector. The conditions were 300 Btu/hr/sq ft input flux level, 7 mph wind, 2 glasses of ⅛" green glass with 88% transmission level, 10 lb/hr/sq ft water flow rate, 80°F ambient, and plain brushed on black paint. The results, considered to be good, were as follows.

Inlet Water Temperature, °F	Efficiency, %
80	70
100	67
140	56
200	38

ZOMEWORKS SKYLID

Skylids are insulated louvers which are placed inside a building behind or beneath skylights, glass roofs, clerestories or vertical windows. They open during sunny weather and close by themselves during very cloudy periods and at night. When the skylids are closed they are an effective thermal barrier, greatly reducing heat losses through glazed openings, thus allowing one to have large glass areas which let solar heat and light into the building during the day without having large heat losses through these glass areas at night. Suggested installations are illustrated in Figures 11.7 and 11.8 as well as in Figure 3.17 above. South is to the left in each of these figures.

FIGURE 11.7: SKYLID—SAW TOOTH ROOF

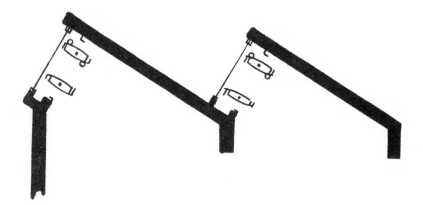

Source: Zomeworks Corporation, Albuquerque, N.M.

FIGURE 11.8: SKYLID—CLERESTORY

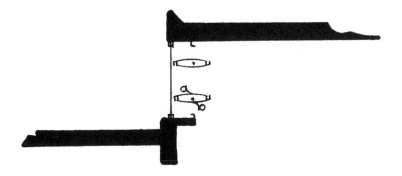

Source: Zomeworks Corporation, Albuquerque, N.M.

Skylids have the additional property of allowing one to regulate, by a manual override, the amounts of heat and light which a skylight admits into a room. One can install a large skylight with confidence that during the summer the room can be prevented from becoming uncomfortably bright or warm. With skylids the same room can be bright and radiantly warm in the winter and dark and cool in the summer.

The essence of the skylid is its simplicity. The louvers move themselves by the shifting weight of the freon. The skylid powers itself: opening when the outside cannister is warmer than the inside and closing when the inside cannister is warmer. The skylid requires no electricity or outside power. For best results louvers should run east-west on windows or skylights sloping 15° to 60°.

Skylids come in 6' to 10' lengths with a width of 22"/panel. Skylids are designed to be installed in finished openings. The finished opening dimensions should be 2" longer in each direction than the total panel length or width. Skylids can be ganged together in units of two or three.

The first skylids have been in operation for two years, opening and closing faithfully. All materials are sturdy and durable. The polyester cloth seals will eventually have to be replaced, but the louvers themselves should continue to function trouble free. Their construction consists of an aluminum skin curved over wooden ribs, an air foil cross section 2" thick at edges, 5" at middle. It rotates on $\frac{5}{16}$" guarded SKF ball bearings. It has glass fiber insulation, a special polyester tube seal, and weighs less than 2 psf. As a cost example a unit 68" x 122" consisting of three panels as of May 1974 costs $340 FOB Albuquerque, N.M.

COMPANY ADDRESSES

AAI Corporation
P.O. Box 6767
Baltimore, Md. 21204

Sunworks, Inc.
669 Boston Post Rd.
Guilford, Conn. 06437

PPG Industries
One Gateway Center
Pittsburgh, Pa. 15222

Tranter, Inc.
735 E. Hazel Street
Lansing, Mich. 48909

Revere Copper and Brass Incorporated
Research and Development Center
Rome, N.Y. 13440

Zomeworks Corporation
P.O. Box 712
Albuquerque, N.M. 87108

Names and Addresses

The people listed below attended the Solar Energy for Heating and Cooling of Buildings Workshop in Washington, D.C., March 21-23, 1973. Their names and addresses are included in the event the reader desires additional information on any particular subject. This list is from PB-223 536, *Proceedings of the Solar Heating and Cooling for Buildings Workshop, Held in Washington, D.C., on March 21-23, 1973,* prepared for the National Science Foundation, issued July 1973.

Mr. Harold W. Aarstad
The Rouse Company
Columbia, Maryland 21044
(301) 730-7700

Dr. Redfield Allen
     Workshop Coordinator
Department of Mechanical Engineering
University of Maryland
College Park, Maryland 20742
(301) 454-2409

Mr. Bruce Anderson
Dubin-Mindell-Bloome Associates
312 Park Road
West Hartford, Connecticut

Mr. Steve Baer
Zomeworks
Box 422
Corrales, New Mexico 87048

Professor E. Barber
Yale University
School of Architecture
180 Yoric Street
New Haven, Connecticut 06510
(203) 432-4763

Mr. Herman G. Barkmann
Sun Mountain Design
P. O. Box 1852
Santa Fe, New Mexico 87501
(505) 982-8907

Mr. James W. Barnes
SES Inc.
70 S. Chapel Street
Newark, Delaware 19711

Mr. Gregory B. Barthold
     Manager, Technical Program
Government Market Development Division
Aluminum Company of America
1200 Ring Building
Washington, D. C. 20036

Dr. Charles D. Beach
Westinghouse Georesearch Laboratory
8401 Baseline Road
Boulder, Colorado 80303
(303) 494-4363

Mr. Walter B. Bienert
Dynatherm Corporation
Marble Court
Cockeysville, Maryland 21030

Dr. Karl W. Böer
Institute of Energy Conversion
University of Delaware
Newark, Delaware 19711
(302) 738-1263

Mr. P. B. Bos, Manager, Solar Projects
Energy Projects Group
P. O. Box 92957
Aerospace Corporation
Los Angeles, California 90045
(213) 648-6406

Mr. Royal Buchanan, Director of Research
A.S.H.R.A.E.
United Engineering Center
345 East 47th Street
New York, New York 10017

Mr. James C. Burke
Arthur D. Little Company
35 Acorn Park
Cambridge, Massachusetts 02140

Mr. William Cherry
Code 760
NASA Goddard Space Flight Center
Greenbelt, Maryland 20771

Mr. David L. Christensen, Research Associate
Center for Environmental Studies
University of Alabama
P. O. Box 1247
Huntsville, Alabama 35807
(205) 895-6362

Mr. Robert Cohen
Environmental Research Laboratories, NOAA
U. S. Department of Commerce
Boulder, Colorado 80302
(303) 499-3773

Dr. James Comly
Corporate R&D Center
General Electric Company
P. O. Box 43
Schenectady, New York 12301
(518) 374-2211 x 54442

Mr. J. Corman
Corporate R&D Center
General Electric Company
P. O. Box 43
Schenectady, New York 12301

Mr. Robert Cunningham
Room 3328, GSA Building
19th and F Streets, N. W.
Washington, D.C. 20405
(202) 343-4048

Dr. Jesse C. Denton
National Center for Energy Management
and Power
113 Towne Building
University of Pennsylvania
Philadelphia, Pennsylvania 19104
(215) 594-5122, 5123

Mr. Fred S. Dubin, P.E.
Dubin-Mindell-Bloome Associates
42 West 39th Street
New York, New York 10018
(212) 868-9700

Prof. John Duffie
Solar Energy Laboratory
Engineering Research Building
1500 Johnson Drive
University of Wisconsin
Madison, Wisconsin 53706

Mr. James Eibling
Battelle Memorial Institute
505 King Avenue
Columbus, Ohio 43021
(614) 299-3151

Mr. Benjamin N. Evans,
     Assistant Director
BRAB-Building Research Institute
2101 Constitution Avenue
Washington, D. C. 20037

Mr. William E. Evans, Teaching Assistant
Environmental Technologies
Yale University, School of Architecture
New Haven, Connecticut 06510
(203) 432-4763

Mr. Frank H. Faust, P. E.
     Consulting Engineer
212 Daleview Lane
Louisville, Kentucky 40207
(Representing A.S.H.R.A.E.)

Mr. Harry C. Fischer
Fischer Associates
Box 1387
Easton, Maryland 21601
(301) 822-4350

Mr. William Fleming
    Supervisor, Market Development
Owens-Corning Fiberglas Corporation
Fiberglas Tower
Toledo, Ohio 43659
(419) 259-3000

Mr. Mike Fortune
Ford Foundation Energy Policy Study
RFF
1776 Massachusetts Avenue, N. W.
Washington, D. C. 20036

Mr. Donald H. Frieling
Battelle Memorial Institute
505 King Avenue
Columbus, Ohio 43220

Mr. Robert E. Gerlach
Helio Associates, Inc.
8230 E. Broadway
Tucson, Arizona 85710
(602) 886-5367

Mr. Oliver DeP. Gildersleeve, Jr.
Philadelphia Electric Company
2301 Market Street-S-10-1
Philadelphia, Pennsylvania 19101

Mr. Herbert T. Gilkey
    Assistant to Managing Director
Air Conditioning and Refrigeration Inst.
1815 North Ft. Myer Drive
Arlington, Virginia 22209
(703) 524-8800

Mr. Ron Griffiths
Jet Propulsion Laboratory
4800 Oak Grove Drive
Pasadena, California 91103
(213) 354-4639

Mr. Keith Gross
104 Prospect Street
New Haven, Connecticut 06511

Dr. R. Grot
Center for Environment
School of Engineering
Princeton University
Princeton, New Jersey

Mr. Blair Hamilton
Educational Facilities Labs
477 Madison Avenue
New York, New York 10022

Dr. W. B. Harris
Texas A&M University
Chemical Engineering Department
College Station, Texas 77843
(713) 846-5810

Mr. Harold Hay
Sky Therm Processes and Engineering
945 Wilshire Boulevard

Dr. James Hill
Center for Building Technology
National Bureau of Standards
Washington, D. C. 20234
(301) 921-3503

Dr. Kenneth E. Johnson
    Associate Director
Center for Environmental Studies
The University of Alabama in Huntsville
P. O. Box 1247
Huntsville, Alabama 35807

Mr. Ralph J. Johnson
    Staff Vice President
National Association of Home Builders
    Research Foundation, Inc.
Box 1627
Rockville, Maryland 20850
(301) 762-4200

Dr. R. C. Jordan
Department of Mechanical Engineering
University of Minnesota
Minneapolis, Minnesota 55455

Dr. Powell Joyner
Trane Company
La Crosse, Wisconsin 54601

Dr. Tamami Kusuda
Center for Building Technology
National Bureau of Standards
Washington, D. C. 20234

Mr. R. Lepson, Senior Engineer
Gas Engineering
Baltimore Gas & Electric Company
Baltimore, Maryland 21203

Dr. Edward Levy
National Academy of Engineering
2101 Constitution Avenue, N. W.
Washington, D. C.

Mr. Leo Loeb
  Manager, Adv. Technology Programs
General Electric Company
Appliance Park
Louisville, Kentucky 40225

Dr. George Löf
  Professor of Civil Engineering
Colorado State University
158 Fillmore Street
Suite 204
Denver, Colorado 80206
(303) 491-8632 or 322-0446

Dr. Harold Lorsch
National Center for Energy Management
  and Power
113 Towne Building
University of Pennsylvania
Philadelphia, Pennsylvania 19104

Mr. Carl R. Maag
Comsat Labs
P. O. Box 115
Clarksburg, Maryland 20734

Mr. John W. Markert
General Services Administration
Room 3036
18th and F Streets, N. W.
Washington, D. C. 20405

Dr. Marshall E. Merriam
  Associate Prof. of Engineering
Materials Science and Engineering Dept.
College of Engineering
University of California
Berkeley, California 94720

Mr. Richard H. Merrick
  Senior Development Engineer
Arkla Air Conditioning Company
Box 534
Evansville, Indiana 47703
(812) 424-3331

Mr. W. C. Moore, Vice President
  & Director of Engineering and Research
York Division
Borg-Warner Corporation
P. O. Box 1592
York, Pennsylvania 17405
(717) 843-0731

Mr. David Namkoong
NASA-Lewis Research Center
21000 Brookpark Road
Cleveland, Ohio 44135
(216) 433-4000 x 6845

Mr. Alwin B. Newton, Vice President
York Division
Borg-Warner Corporation
136 Shelbourne Drive
York, Pennsylvania 17403
(717) 843-0731 x 223

Mr. William Paynton
  Manager, Government Products
Texas Instruments
Attleboro, Massachusetts
(617) 222-2800

Mr. David Rabenhorst
Applied Physics Laboratory
8621 Georgia Avenue
Silver Spring, Maryland 20910
(301) 953-7100

Mr. P. Richard Rittlemann
Burt, Hill and Associates
610 Mellon Bank Building
Butler, Pennsylvania 16001
(412) 285-4761

Mr. Frank Rom
NASA-Lewis Research Center
21000 Brookpark Road
Cleveland, Ohio 44135
(216) 433-4000 x 6266

Dr. Jeffrey H. Rumbaugh
  Senior Staff Engineer
Office of Vice President
Potomac Electric Power Company
1900 Pennsylvania Avenue, N. W.
Washington, D. C. 20006
(202) 872-2714

Mr. T. Santala
Texas Instruments, Inc.
34 Forest Street
Attleboro, Massachusetts 02703

Dr. Stephen L. Sargent
Department of Mechanical Engineering
University of Maryland
College Park, Maryland 20740
(301) 454-4216

Mr. Roger Schmidt
Systems and Research Center
Honeywell, Incorporated
2700 Ridgway Road
Minneapolis, Minnesota 55413
(612) 331-4141 x 4191

Mr. Richard Schoen
School of Architecture/Urban Planning
UCLA
405 Hilgard Avenue
Los Angeles, California 90024

Dr. C. Sepsy
Ohio State University Research Foundation
1314 Kinnear Road
Columbus, Ohio 43212

Mr. David J. Shamp
Custom Applied Power Corporation
6060 Farrington Avenue
Alexandria, Virginia 22304

Mr. S. Siegel
Oak Ridge National Laboratory
P. O. Box Y
Oak Ridge, Tennessee 37830

Mr. W. Gene Steward
Sugar Loaf Road
Swiss Peaks
Boulder, Colorado 80302
(303) 499-1000 x 3833

Mr. A. J. Streb
Dynatherm Corporation
Marble Ct. & Industry Lane
Cockeysville, Maryland 21030
(301) 668-6570

Mr. Elmer R. Streed
Lockheed Missiles & Space Company
Palo Alto Research Laboratory
Department 52/21 Building 205
3251 Hanover Street
Palo Alto, California 94304

Mr. Robert P. Stromberg
Division 1212
Sandia Laboratory
P. O. Box 5800
Albuquerque, New Mexico 87115
(505) 264-8170

Mr. Herbert H. Swineburne
Nolan and Swineburne Associates
1624 Locust Street
Philadelphia, Pennsylvania 19103
(215) 732-1401

Dr. George C. Szego, President
Inter Technology Corporation
Warrenton, Virginia 22186
(703) 273-5112

Mr. William Peter Teagan
Thermo Electron Corporation
85 First Avenue
Waltham, Massachusetts 02154
(617) 890-8700

Dr. Maria Telkes
Institute of Energy Conversion
University of Delaware
Newark, Delaware 19711
(302) 738-2887

Mr. William R. Terrill
Building B. VFSC
General Electric Company
P. O. Box 8661
Philadelphia, Pennsylvania 19101

Dr. H. E. Thomason
6802 Walker Mill Road, S. E.
Washington, D. C. 20027
(301) 336-4042

Mr. H. J. L. Thomason, Jr.
6802 Walker Mill Rd., S. E.
Washington, D. C. 20027
(301) 336-4042

Mr. George E. Turnbull
Yale University
232 Ellsworth Avenue DF
New Haven, Connecticut 06511
(203) 562-6264

Mr. J. D. Walton
Engineering Experiment Station
Georgia Tech
Atlanta, Georgia 30332

Dr. Jerome Weingart
Environmental Quality Laboratory
California Institute of Technology
Pasadena, California 91109
(213) 795-6841 x 1134, 2783

Mr. E. P. Whitlow
Whirlpool Corporation, R&E
Monte Road
Benton Harbor, Michigan 49022

Mr. John H. Whitney
   Manager, Marketing Services Division
Potomac Electric Power Company
1900 Pennsylvania Avenue, N. W.
Washington, D. C. 20006

Mr. L. Stephen Windheim
   Senior Vice President
Leo A. Daly Company
1025 Connecticut Avenue, N. W.
Washington, D. C. 20036
(202) 265-3113

Dr. John I. Yellott
John Yellott Engineering Associates, Inc.
901 West El Caminito Drive
Phoenix, Arizona 85020
(602) 943-5805

General References

The following references have not been included previously in the text but are suggested as general references for additional reading. The first reference listed, it should be noted, is an extensive and thorough discussion of the theory and calculations involved in low temperature solar thermal processes.

(1) Duffie, J.A. and Beckman, W.A., *Solar Energy Thermal Processes,* John Wiley & Sons, New York, 1974.

(2) Daniels, F., *Direct Use of the Sun's Energy,* Yale University Press, 1964.

(3) Steadman, P., *Energy, Environment and Building,* Cambridge University Press, 1975.

(4) Donovan, P.; Woodward, W.; Cherry, W.P.; Morse, F.H. and Herwig, L.O.; PB-221 659, *An Assessment of Solar Energy as a National Resource,* December 1972.

(5) Berg, C.A., COM-73-10856, *Energy Conservation Through Effective Utilization,* February 1973.

(6) Thomas, R.L., N73-22748, *The Utilization of Solar Energy to Help Meet Our Nation's Energy Needs,* May 1973.

(7) PB-235 483, *Proceedings of the Solar Heating and Cooling for Buildings Workshop, Washington, D.C., March 21-23, 1973, Part II: Panel Sessions, March 23,* April 1974.

ENERGY
FROM SOLID WASTE 1974

by Frederick R. Jackson

Pollution Technology Review No. 8

Energy Technology Review No. 1

The solid waste disposal problem is reaching alarming proportions everywhere. The United States alone produces close to 300 million tons of solid waste per year, which is equivalent to about one ton per person.

At the present time the prevalent methods of disposal are dumping and sanitary landfill. Municipal incineration disposes of a small portion only, with attendant high capital and operating costs.

Many methods have been proposed for coping with the problem, such as source separation, source reduction, or material recovery. However, with the energy crisis descending upon us, producing energy from waste is becoming more and more attractive. An estimate currently making the rounds in financial circles is that when the price of crude oil reaches $7.00 a barrel, alternate sources of energy become practicable.

This book is based primarily upon information from studies conducted under the auspices of the EPA. Its foremost topic is burning of solid wastes to create steam directly. The air pollution problem created by burning can be solved quite easily with known technology. Solid wastes are low in sulfur, consequently there is no SO_2 removal problem.

Another technique that may assume more importance in the future is the controlled pyrolysis of wastes, yielding so-called pyrolysis gas or oil. Chapter seven is devoted to this. The final chapter discusses European practice, which is historically far more extensive than that of the U.S. A condensed table of contents follows here:

ISBN 0-8155-0528-0

163 pages

UTILIZATION OF WASTE HEAT
FROM POWER PLANTS 1974

by David Rimberg

Pollution Technology Review No. 14

Energy Technology Review No. 3

Present-day steam-driven turbo-electric power plants in the United States discharge as waste heat an amount of energy roughly equivalent to twice their total electricity-generating capacity. This energy is most difficult to utilize, because it is degraded in temperature. The effluent water is warm, but it is far from being near the boiling point. It constitutes a necessary, but unwanted, by-product of the energy conversion process for generating electricity.

The growing quantities of waste heat discharged, and the increasing ecological anxieties about undesirable growths, energy utilization and thermal discharge problems, have stimulated an examination of methods for productively using energy presently wasted on the environment.

Part I of this book discusses the reasons for present-day ineffective utilization, but It also shows ways and means to promote more efficient energy usage.

Part II assesses the cause, magnitude, and possible effects of heat discharges into water from steam electric power plants and related condenser cooling systems. Also contained in this section is a discussion of the thermodynamics of the electric power generation cycle detailing the reasons for the "inefficiencies" in by-product heat generation.

Ultimately all the accessible waste energy appears as low temperature heat, and the term "waste heat utilization" refers to the performance of useful functions with this heat before it is discharged into the environment. Within the framework of this book, the minimum temperature of the effluent water considered for subsequent use is about 38°C or 100°F. These uses are discussed in **Part III**: Food production in agriculture, hydroponics and aquaculture (fish farming) and the use of heat in wastewater treatment.

In the past the dissipation of waste heat was accomplished by wet and dry cooling towers and cooling ponds, lakes and streams. These methods are now being challenged by some sectors of society, and industry is being forced to consider their environmental impact. In this regard **Part IV** discusses the research needs necessary to equate the complicated interactions of the physical, engineering, biological, and social aspects of this waste heat problem.

This Pollution Technology Review is based on studies conducted by industrial and engineering firms or university research teams under the auspices of various governmental agencies, e.g. The National Water Commission, Oak Ridge National Laboratory Federal Water Pollution Control Administration, Office of Water Resources Research, Environmental Protection Agency, National Science Foundation, and the Department of Commerce.

A partial and condensed table of contents follows here:

ISBN 0-8155-0555-8

175 pages

RECYCLING AND RECLAIMING
OF MUNICIPAL SOLID WASTES 1975

by F. R. Jackson

Pollution Technology Review No. 17

More and more cities, large and small, are exploring new technological systems for recycling solid municipal wastes.

Naturally the greatest interest centers upon those recycling systems that permit profitable reclamation of saleable items and materials from the town's refuse.

Articles made of metal can be cleaned, washed, compressed, and sold to the scrap metals industry. Glass can be comminuted and even separated as to color.

Paper and cardboard is a promising target for recycling, because paper fiber is the largest single component of municipal wastes and potentially very valuable.

Water-insoluble organic matter, such as plastics and wood, can be reclaimed and processed for conversion into energy in a variety of ways.

As these new technologies are developed, they will add to the alternatives available to a community, when it decides to recycle some of its wastes as part of its total solid waste management program. Numerous alternate ways and means are needed, because local conditions vary widely, and no single line of approach is capable of meeting every community's needs.

This Pollution Technology Review surveys information available up to the middle of 1974, and is based primarily upon studies conducted by industrial and engineering firms under the auspices of the EPA and other government agencies. Forward-thinking municipal administrators will find this book most useful, because it collates comprehensively the results of governmental and industrial research.

A partial and condensed table of contents follows here.

ISBN 0-8155-0560-4

342 pages

INCINERATION OF SOLID WASTES 1974

by Fred N. Rubel

Pollution Technology Review No. 13

This book gives advanced technical and economic information to promote efficient incineration of municipal refuse, and will help to erase the bad neighbor image of incinerator installations within the community. Special incinerator applications, largely outside the scope of municipal waste disposal, have not been neglected.

Incineration offers the most significant volume reduction of solid organic wastes when compared with all other disposal methods.

The most undesirable side effect is, or rather was, air pollution, and here the primary concern is with emissions of particulates, rather than with gases and odors. Many sophisticated devices for nearly complete retention of these solid particles have been developed in recent years, and special emphasis has been placed upon them in the descriptions of apparatus.

Most of the data and other information presented in this book are based upon detailed reports on government contracts fulfilled by industrial companies under the auspices of the Environmental Protection Agency.

A partial and condensed table of contents follows here. Chapter headings are given, followed by examples of important subtitles.

ISBN 0-8155-0551-5

246 pages

LARGE SCALE COMPOSTING 1974

by M. J. Satriana

Pollution Technology Review No. 12

Composting is one of the oldest solid waste disposal methods known to man. This microbiological treatment process has the capability of converting municipal refuse and other organic waste solids into a product which has a lower bulk than the original waste, is stable, and can be recycled into the ecological sphere.

Aerobic, thermophilic composting is generally believed to provide the most rapid and complete decomposition of oxidizable or putrescible organic matter. It is especially applicable to decomposing material consisting primarily of discrete organic solids with open pore spaces.

Rough-quality compost, cheaply produced, has real potential for the reclamation of poor and spoiled lands (as from strip mining) providing it can be transported there cheaply (e.g. by barges).

This book, based primarily upon reports produced under the auspices of the Environmental Protection Agency, presents an overview existing large-scale composting methods, both here and abroad. It compares costs and partial recovery of these costs, but retains a significant emphasis on the workable disposal of urban waste versus other and usually more expensive methods of getting rid of solid organic refuse. Nine chapters. A partial and condensed table of contents follows here.

ISBN 0-8155-0548-3

269 pages

SANITARY LANDFILL TECHNOLOGY
1974

by Samuel Weiss

Pollution Technology Review No. 10

This Pollution Technology Control Review surveys all the latest technical information on this subject. It is based primarily on studies conducted by industrial or engineering firms, under the auspices of the Environmental Protection Agency.

Sanitary landfill practice is an engineering method of disposing of solid wastes on land, whether they are of municipal or industrial origin. There is no resemblance to the old-fashioned garbage and rubbish dump. There are no fires, no obnoxious fumes or smoke, no flies and no rodents or other scavengers.

As stated in the introduction to this book, a sanitary landfill is not only an acceptable and economic method of solid waste disposal, it provides also an excellent way to improve the commercial value of otherwise unsuitable or marginal land areas within a few years.

A partial and condensed table of contents follows here.

ISBN 0-8155-0542-6

300 pages

ENERGY IN THE CITY
ENVIRONMENT 1973

Edited by Robert N. Rickles

This book is based upon material supplied by
Robert Rickles, the Institute for Public Transportation, and the New York Board of Trade.

I: CRISES IN ENERGY SOURCES AND PRODUCTION
1. OVERVIEW (1972-1985)
   Primary Reliance on Conventional
     Fuel Sources
   Problems in Research and Technology
   Lead Time for Breeder Reactors
   Fusion
   Solar Energy
   Coal Gasification

2. OIL
   NEPA and Other Environmental Issues
   Accelerated Government Leasing
     Programs on Outer Continental
     Shelf
   Oil Imports
   Costs and Risks: Economic, National
     Security and Environmental
     Problems
   Improved Exploration Incentives
   Product Prices versus Foreign Imports
   Tax Incentives

3. GAS
   Accelerated Government Leasing of
     Outer Continental Shelf
   Amendments to Natural Gas Act
   Security of Contract Terms
   Competitive Pricing
   LNG/SNG — Imported and Domestic
   Interstate and Intrastate Marketing
     Policies

4. ELECTRICITY
   Generating Plant Licensing and
     Regulations
   Nuclear Technology
   Pumped Storage
   Availability of Low Sulfur Fuel Oil
     and Gas
   Efficiency of Conventional Sources
   Purchasing Power (Imports)

II: CRISES IN ENERGY DEMANDS
1. OVERVIEW
   Are Our Priorities and Objectives
     Compatible and Ready for Action?
   Competition for Available Supplies:
     National and International
     Demands Constrain City Energy
     Resources
   Effects of Differing Demand Break-
     downs
   Institutional Alternatives for
     Energy Management

2. REALISTIC APPRAISAL OF
   DEMAND GROWTH RATES (through
   1985)
   Population Growth (Inevitable
     Modest Demand Expansion)
   Social Side Effects of Energy Con-
     straint Policy

   Additional Energy of Environmental
     Protection
   Air Pollution and Energy

3. CONSERVATION OF ENERGY
   RESOURCES
   Rationing and End Use Controls
   Taxing Certain Energy Uses
   Changes in Life Style
   Improved Energy Efficiency
   New Buildings
   Mass Transit
   Appliances

**III: ELEMENTS IN THE ENERGY-
ENVIRONMENT BALANCE**
1. OVERVIEW
   Need to Develop New Energy Reserves
     Without Unreasonable Environ-
     mental Delays
   Ten Year Planning Lead
   New Explorations to Preclude Shortages

2. IMPACT OF INADEQUATE ENERGY
   SUPPLIES
   Economy-GNP-Unemployment-
     Inhibition of Investments
   Insecurity and Reduced Standard of
     Living
   Slowdown and Abandonment of Public
     Sector Programs
   National Security
   Environmental Protection: Air Pollu-
     tion, Water and Waste Management

3. ENERGY-ENVIRONMENT BALANCING
   Air Pollution
   Offshore Drilling
   Increased Tanker Traffic
   Reliance on Fuel Imports
   Oil and LNG
   Nuclear Power
   Radiation and Waste Disposal
   Interim Nuclear Licensing
   Selection and Nature of Plant Sites
   Air Quality
   Thermal Pollution

IV. NYS POWER PLANT SITING ACT

The important **charts** in the **appendix**
present a list of energy conservation measures for the short-term (1972-1975), mid-
term (1976-1980) and the long-term (beyond 1980) for the Transportation, Resi-
dential/Commercial, Industrial and Electric
Utility sectors. The charts also indicate
estimated maximum attainable energy savings, possible means for implementing each
conservation measure, and pros and cons.

ISBN 0-8155-5019-7

173 pages

ENVIRONMENTAL SOURCES
AND EMISSIONS HANDBOOK 1975

by Marshall Sittig

Environmental Technology Handbook No. 2

Environmental pollution with its far-reaching effects is only now beginning to be understood. In this practical handbook the various modifications a given pollutant can assume, are traced back to their origins, and the intermedia transfers are discussed fully. Intermedia transfers include direct transfer (removal of a pollutant from one medium and its disposal in another) or indirect (pollution created in another medium and usually in another form by a basic change in a process or industry).

Also, an example of insidious conversion is mercury. Metallic mercury is relatively innocuous, but when organic molecules or dead organisms are present in rivers and lakes, mercury reacts with these molecules to form toxic methylmercury compounds which are excreted very slowly by fish and man.

The implications are formidable: Chemical, biochemical, technological, statistical, mineralogical, zoological, pharmacological, medical, legal, legislative and public health involvements are all too obvious.

This volume therefore surveys the origins of both air and water pollution. Significant emphasis is placed on altered pollution which can result when an air pollutant is intentionally or accidentally transferred to an aqueous stream or vice versa.

Sometimes pollutants react with each other or initiate deleterious chain reactions. Yet this same reactivity may be the key to efficient removal: by adding certain chemicals or flocculants or microorganisms, toxic substances may be carried off physically or converted not only to nontoxic substances, but also into useful products.

The why and where is in this book, commensurate with present-day technology, without becoming too sophisticated or losing sight of the all important economic considerations.

Operators or potential operators of processes which produce pollutants will find this volume quite useful. Besides discussing all sorts of pollution sources, it should help to define industry-wide emission practices and magnitudes.

A partial and condensed table of contents follows here. The book contains 200 subject entries and 382 tables with figures and graphs. Descriptions are mostly based on studies conducted by industrial and engineering firms or university research teams under the auspices of various government agencies. As is the case with other handbooks in this series the entries are arranged in an alphabetical and encyclopedic fashion:

Aluminum Production, Primary
Aluminum Industry, Secondary
Aluminum Sulfate Manufacture
Ammonia Manufacture
Ammonium Nitrate Manufacture
Ammonium Sulfate Production
Asbestos Products Industry
Asphalt Roofing Manufacture
Asphaltic Concrete Plants
   (Asphalt Batching)
Automobile Body Incineration
Automotive Vehicle Operation
Bauxite Refining
Beet Sugar Manufacture
Boat & Ship Operation
Brass & Bronze Ingot Prod.
   (Secondary Copper Industry)
Brick & Related Clay
   Products Manufacture
Builders Paper Manufacture
Calcium Carbide Manufacture
Calcium Chloride Manufacture
Carbon Black Manufacture
Castable Refractory Manufacture
Cement Manufacture
Ceramic Clay Manufacture
Charcoal Manufacture
Chlor-Alkali Manufacture
Clay & Fly Ash Sintering
Coal Cleaning
Coal Combustion, Anthracite
Coal Combustion, Bituminous
Coffee Roasting
Coke (Metallurgical) Manufacture
Combustion Sources
Concrete Batching
Conical Burners
Copper (Primary) Smelting
Cotton Ginning
Dairy Industry
Dry Cleaning
Electroplating
Explosives Manufacture
Feedlots
Fermentation Industry
Ferroalloy Production
Fertilizer Industry
Fiberglass Manufacture
Fish Processing
Frit Manufacture
Fruit & Vegetable Industry
Fuel Oil Combustion
Glass Manufacture
Gold & Silver Mining & Production
Grain & Feed Mills & Elevators
Gypsum Manufacture
Hydrochloric Acid Manufacture
Hydrofluoric Acid Manufacture
Hydrogen Peroxide Manufacture
Incineration, Municipal Refuse
Inorganic Chemical Industry
Iron Foundries
Iron & Steel Mills
Lead Smelting, Primary
Lead Smelting, Secondary
Leather Industry
Lime Manufacture
Liquefied Petroleum Gas Combustion
Magnesium Smelting, Secondary

Meat Industry
Mercury Mining & Production
Mineral Wool Manufacturing
Natural Gas Combustion
Nitric Acid Manufacture
Nitrogen Fertilizer Manufacture
Nonferrous Metals Industry
Open Burning
Orchard Heating
Organic Chemicals Manufacture
Paint & Varnish Manufacture
Paperboard Manufacture
Perlite Manufacture
Pesticide Manufacture
Petroleum Marketing & Transportation
Petroleum Refining
Petroleum Storage
Phosphate Chemicals Manufacture
Phosphate Fertilizer Manufacture
Phosphate Rock Processing
Phosphoric Acid Manufacture
Phosphorus Manufacture
Phosphorus Oxychloride Manufacture
Phosphorus Pentasulfide Manufacture
Phosphorus Pentoxide Manufacture
Phosphorus Trichloride Manufacture
Phthalic Anhydride Manufacture
Plastics Industry
Potassium Chloride Production
Potassium Dichromate Manufacture
Potassium Sulfate Manufacture
Poultry Processing
Power Plants
Printing Ink Manufacture
Pulp & Paper Industry
Railway Operation
Refractory Metal (Mo, W) Production
Rubber Industry, Synthetic
Sand & Gravel Processing
Seafood Processing Industry
Sewage Sludge Incineration
Soap & Detergent Manufacture
Sodium Bicarbonate Manufacture
Sodium Carbonate Manufacture
Sodium Chloride Manufacture
Sodium Dichromate Manufacture
Sodium Metal Manufacture
Sodium Silicate Manufacture
Sodium Sulfite Manufacture
Starch Manufacture
Stationary Engine Operation
Steel Foundries
Stone Quarrying & Processing
Sugar Cane Processing
Sulfur Production
Sulfuric Acid Manufacture
Surface Coating Application
Synthetic Fiber Manufacture
Terephthalic Acid Manufacture
Textile Industry
Timber Industry
Tire & Inner Tube Manufacture
Titanium Dioxide Manufacture
Urea Manufacture
Wood Veneer & Plywood Products Industry
Wood Waste Combustion
Zinc Processing, Secondary
Zinc Smelting, Primary

ISBN 0-8155-0568-X

521 pages